Uwe Sander

# Numerical and Algebraic Studies for the Control of Quantum Systems

Uwe Sander

# Numerical and Algebraic Studies for the Control of Quantum Systems

Controllability and Optimal Control in Finite-Dimensional Quantum Systems

Südwestdeutscher Verlag für Hochschulschriften

**Impressum/Imprint (nur für Deutschland/ only for Germany)**
Bibliografische Information der Deutschen Nationalbibliothek: Die Deutsche Nationalbibliothek verzeichnet diese Publikation in der Deutschen Nationalbibliografie; detaillierte bibliografische Daten sind im Internet über http://dnb.d-nb.de abrufbar.

Alle in diesem Buch genannten Marken und Produktnamen unterliegen warenzeichen-, marken- oder patentrechtlichem Schutz bzw. sind Warenzeichen oder eingetragene Warenzeichen der jeweiligen Inhaber. Die Wiedergabe von Marken, Produktnamen, Gebrauchsnamen, Handelsnamen, Warenbezeichnungen u.s.w. in diesem Werk berechtigt auch ohne besondere Kennzeichnung nicht zu der Annahme, dass solche Namen im Sinne der Warenzeichen- und Markenschutzgesetzgebung als frei zu betrachten wären und daher von jedermann benutzt werden dürften.

Verlag: Südwestdeutscher Verlag für Hochschulschriften GmbH & Co. KG
Dudweiler Landstr. 99, 66123 Saarbrücken, Deutschland
Telefon +49 681 37 20 271-1, Telefax +49 681 37 20 271-0
Email: info@svh-verlag.de
Zugl.: München, TU, Diss., 2010

Herstellung in Deutschland:
Schaltungsdienst Lange o.H.G., Berlin
Books on Demand GmbH, Norderstedt
Reha GmbH, Saarbrücken
Amazon Distribution GmbH, Leipzig
ISBN: 978-3-8381-2484-1

**Imprint (only for USA, GB)**
Bibliographic information published by the Deutsche Nationalbibliothek: The Deutsche Nationalbibliothek lists this publication in the Deutsche Nationalbibliografie; detailed bibliographic data are available in the Internet at http://dnb.d-nb.de.

Any brand names and product names mentioned in this book are subject to trademark, brand or patent protection and are trademarks or registered trademarks of their respective holders. The use of brand names, product names, common names, trade names, product descriptions etc. even without a particular marking in this works is in no way to be construed to mean that such names may be regarded as unrestricted in respect of trademark and brand protection legislation and could thus be used by anyone.

Publisher: Südwestdeutscher Verlag für Hochschulschriften GmbH & Co. KG
Dudweiler Landstr. 99, 66123 Saarbrücken, Germany
Phone +49 681 37 20 271-1, Fax +49 681 37 20 271-0
Email: info@svh-verlag.de

Printed in the U.S.A.
Printed in the U.K. by (see last page)
ISBN: 978-3-8381-2484-1

Copyright © 2011 by the author and Südwestdeutscher Verlag für Hochschulschriften GmbH & Co. KG and licensors
All rights reserved. Saarbrücken 2011

To my parents.

# Abstract

In this thesis, two aspects of control theory, namely controllability and optimal control, are applied to quantum systems. The presented results are based on group theoretical techniques and numerical studies. By Lie-algebraic analysis, the controllability properties of systems with an arbitrary topology are described and related to the symmetries existing in these systems. We find that symmetry precludes full controllability. Our work investigates well-known control systems and gives rules for the design of new systems. Furthermore, theoretical and numerical concepts are instrumental to studying quantum channels: Their capacities are optimised using gradient flows on the unitary group in order to find counterexamples to a long-established additivity conjecture. The last part of this thesis presents and benchmarks a modular optimal control algorithm known as GRAPE. Numerical tests show how the interplay of its modules can be optimised for higher performance, and how the algorithm performs in comparison to a Krotov-type optimal control algorithm. It is found that GRAPE performs particularly well when aiming for high qualities.

# Zusammenfassung

In der vorliegenden Arbeit werden zwei Aspekte der Kontrolltheorie, Kontrollierbarkeit und Optimale Kontrolle, auf Quantensysteme angewandt. Die gefundenen Resultate basieren dabei auf gruppentheoretischen Methoden und numerischen Untersuchungen. Mit Hilfe Lie-algebraischer Analysen werden die Kontrollierbarkeitseigenschaften von Spinsystemen beliebiger Topologie beschrieben und in Beziehung gesetzt zu in den Systemen vorhandenen Symmetrien. Es stellt sich heraus, dass volle Kontrollierbarkeit nicht erreicht werden kann, wenn eine Symmetrie im System vorhanden ist. Die Arbeit analysiert bekannte Kontrollsysteme und stellt Regeln für den Entwurf neuer Systeme auf. Desweiteren kommen theoretische und numerische Konzepte auch bei der Betrachtung von Quantenkanälen zum Einsatz: Hier werden Kanalkapazitäten durch Gradientenflüsse auf der unitären Gruppe optimiert, um Gegenbeispiele für eine bekannte Additivitätsvermutung zu finden. Im letzten Teil der Arbeit wird ein modularer Algorithmus zur optimalen Steuerung von Quantensystemen vorgestellt, der als GRAPE-Algorithmus bekannt ist. Seine Module werden analysiert und mit Hilfe numerischer Tests optimiert, so dass ein umfassender Benchmark entsteht. Ein Leistungsvergleich mit einem anderen, auf Krotov zurückgehenden Algorithmus schließt die Arbeit ab. In den meisten Tests kann GRAPE dabei hohe Qualitäten schneller erreichen als der Krotov-basierte Algorithmus.

# Contents

1 **Introduction**    1
     1.1   Motivation and overview . . . . . . . . . . . . . . . . . . . . . . 1
     1.2   Organisation of this thesis . . . . . . . . . . . . . . . . . . . . . 3

2 **An introduction to some definitions and concepts of group theory**    5
     2.1   Introduction . . . . . . . . . . . . . . . . . . . . . . . . . . . . . 5
     2.2   Definitions . . . . . . . . . . . . . . . . . . . . . . . . . . . . . . 5
     2.3   Matrix representation . . . . . . . . . . . . . . . . . . . . . . . . 7
     2.4   Lie groups and algebras . . . . . . . . . . . . . . . . . . . . . . 8
         2.4.1   General . . . . . . . . . . . . . . . . . . . . . . . . . . . . 8
         2.4.2   Pauli operator basis . . . . . . . . . . . . . . . . . . . . . 10
         2.4.3   Lie subgroups . . . . . . . . . . . . . . . . . . . . . . . . 11

3 **Controllability and symmetry in spin systems**    13
     3.1   Introduction . . . . . . . . . . . . . . . . . . . . . . . . . . . . . 13
         3.1.1   Overview . . . . . . . . . . . . . . . . . . . . . . . . . . . 13
         3.1.2   Organisation and main results . . . . . . . . . . . . . . 14
     3.2   Quantum dynamical control systems . . . . . . . . . . . . . . . 15
     3.3   Full controllability and symmetry-restricted controllability in closed quantum systems . . . . . . . . . . . . . . . . . . . . . . . . . . . . . . . . 17
         3.3.1   Algorithm for computing the Lie closure . . . . . . . . 17
         3.3.2   Notation: coupling graphs and interactions . . . . . . . 18

|  |  | 3.3.3 Characterisation by symmetry and antisymmetry | 19 |
|---|---|---|---|
|  | 3.4 | Introductory examples with symmetry-restricted controllability | 22 |
|  |  | 3.4.1 Example 1: joint $S_2$ symmetry | 22 |
|  |  | 3.4.2 Example 2: individual permutation symmetry | 23 |
|  | 3.5 | Task controllability | 24 |
|  | 3.6 | Discussion of inner and outer symmetries | 26 |
|  |  | 3.6.1 Systems with outer symmetry | 26 |
|  |  | 3.6.2 Systems with inner symmetry | 30 |
|  |  | 3.6.3 Qubit chains with minimalistic controls | 34 |
|  | 3.7 | Absence of symmetry versus full controllability | 37 |
|  |  | 3.7.1 Absence of symmetry implies (semi-)simplicity | 38 |
|  |  | 3.7.2 Conditions for simplicity | 38 |
|  |  | 3.7.3 Sufficient conditions for full controllability | 39 |
|  |  | 3.7.4 Necessary conditions for full controllability | 44 |
|  | 3.8 | Efficient controllability | 45 |
|  | 3.9 | Conclusions | 45 |
| 4 | **Numerical studies on the additivity of quantum channel capacities** | | **47** |
|  | 4.1 | Introduction | 47 |
|  | 4.2 | Quantum channels | 48 |
|  |  | 4.2.1 Definition | 48 |
|  |  | 4.2.2 The Bloch sphere representation | 49 |
|  |  | 4.2.3 Examples | 49 |
|  | 4.3 | The additivity conjecture | 51 |
|  | 4.4 | The Werner-Holevo channel as a counterexample | 52 |
|  | 4.5 | Optimising the output purity using gradient flows | 53 |
|  |  | 4.5.1 Statement of the problem | 53 |
|  |  | 4.5.2 Description of the numerical procedure | 54 |

|  |  | 4.5.3 Results and further developments . . . . . . . . . . . . . . . . . . . . | 56 |

## 5 Benchmarking a concurrent-update optimal-control algorithm — 61

- 5.1 Introduction . . . . . . . . . . . . . . . . . . . . . . . . . . . . . . . . . . . . . 61
  - 5.1.1 Overview . . . . . . . . . . . . . . . . . . . . . . . . . . . . . . . . . . 61
  - 5.1.2 Organisation . . . . . . . . . . . . . . . . . . . . . . . . . . . . . . . . 62
- 5.2 The optimal control framework . . . . . . . . . . . . . . . . . . . . . . . . . 62
  - 5.2.1 Pontryagin's maximum principle . . . . . . . . . . . . . . . . . . . . 63
- 5.3 Optimal control for quantum systems . . . . . . . . . . . . . . . . . . . . . 64
  - 5.3.1 The quality function . . . . . . . . . . . . . . . . . . . . . . . . . . . 65
- 5.4 Algorithmic scheme . . . . . . . . . . . . . . . . . . . . . . . . . . . . . . . . 66
- 5.5 Gradient computation . . . . . . . . . . . . . . . . . . . . . . . . . . . . . . 67
  - 5.5.1 The gradient formula with respect to the control amplitudes . . . . 67
  - 5.5.2 The gradient formula with respect to time . . . . . . . . . . . . . . 72
  - 5.5.3 The gradient formula with respect to phase . . . . . . . . . . . . . 73
- 5.6 Update methods . . . . . . . . . . . . . . . . . . . . . . . . . . . . . . . . . . 74
  - 5.6.1 Steepest ascent . . . . . . . . . . . . . . . . . . . . . . . . . . . . . . 74
  - 5.6.2 Newton and quasi-Newton methods . . . . . . . . . . . . . . . . . 74
  - 5.6.3 Conjugate gradients . . . . . . . . . . . . . . . . . . . . . . . . . . . 75
  - 5.6.4 Line search . . . . . . . . . . . . . . . . . . . . . . . . . . . . . . . . . 76
- 5.7 Comparison with a sequential-update algorithm . . . . . . . . . . . . . . . 76
- 5.8 Hybrid algorithms . . . . . . . . . . . . . . . . . . . . . . . . . . . . . . . . . 79
- 5.9 Numerical studies . . . . . . . . . . . . . . . . . . . . . . . . . . . . . . . . . 79
  - 5.9.1 Test environment . . . . . . . . . . . . . . . . . . . . . . . . . . . . . 79
  - 5.9.2 Toy models used for numerical optimisations . . . . . . . . . . . . 80
  - 5.9.3 Comparison of gradient methods . . . . . . . . . . . . . . . . . . . 82
  - 5.9.4 Comparison of update methods . . . . . . . . . . . . . . . . . . . . 88
  - 5.9.5 Comparison of sequential- and concurrent-update algorithms . . . 90

CONTENTS

    5.10 Conclusions . . . . . . . . . . . . . . . . . . . . . . . . . . . . . . . 98

**6 Conclusions and outlook**    **101**

**A The superoperator formalism 101**    **105**

**B Proof of Lemma 4**    **107**

**C Comparison of gradient methods**    **109**

**D The interior-point algorithm**    **113**

**E Additional numerical results**    **115**

**Bibliography**    **121**

CHAPTER 1

# Introduction

> Science is what we understand well
> enough to explain to a computer. Art is
> everything else we do.
>
> Donald E. Knuth

## 1.1 Motivation and overview

Death, the weather, tax refunds - there are things that no one can control. Control theory is about everything else. It is a field of research that has arisen from control engineering and is nowadays a subfield of applied mathematics [1]. Its main concerns are dynamical systems that are influenced by external variables. This general setup makes it a truly interdisciplinary field with applications in many different areas, such as engineering, finance, economics, and physics. Typical questions one wishes to answer by control theory are: Is the system stable? How can the system be steered to show a desired behaviour? What are the possible outputs of the system?

These questions also appear naturally in quantum information science, where one studies the processing of information using quantum mechanical systems [2]. In fact, the ability to control the state and the dynamics of complex quantum systems is one of the main goals in this field. In the last two decades, quantum information theory has attracted growing interest and has become an area of lively research, connecting physics, mathematics, and computer science. Its scope ranges from fundamental questions about quantum mechanics to technological applications like quantum cryptography [3] and quantum computation. A model for the latter was proposed by Feynman [4] almost thirty years ago: His idea of

CHAPTER 1: INTRODUCTION

using a quantum computer for simulating other quantum systems or for solving complex problems has sparked a tremendous research effort to realise a universal large-scale quantum computer. Alongside the experimental research, algorithms have been developed that exploit the power of quantum computers to solve problems that are hard to solve on a classical computer. The first prominent example of such a quantum algorithm was the Deutsch algorithm [5, 6], which showed the superiority of quantum computers for a certain problem but was of little practical use. In contrast, Shor's algorithm [7] to factorise large numbers in polynomial time is of huge practical relevance for classical cryptography.

For the purpose of large-scale quantum computation, coherently controlled systems are needed. The number of these systems that can profit from optimal control has risen in the past few years [8, 9, 10, 11, 12, 13, 14]. Optimal control theory (OCT) generally deals with methods for finding control variables that minimise or maximise a given performance index [15, 16]. This requires the formulation of the control objective in mathematical terms, usually in the form of a cost functional. In many cases, constraints have to be included into the problem, typically as equalities or inequalities, such that the resulting problem can be described as: find the values of the control variables that optimise the objective while respecting the constraints. As outlined by to Sussmann and Willems [17], OCT originated from the calculus of variations, in particular in curve minimisation problems that were studied by Johann Bernoulli and others at the end of the 17th century. Modern OCT is the result of a long process that culminated in the work of Pontryagin [18] and Bellman [19] who established conditions for optimality in the 1950s. Pushed by ongoing technological advancements, physicists have become more and more interested in the coherent control of quantum systems, in particular for quantum information processing. As these systems are sensitive to noise, an important question of quantum control is how to implement a given operation on the system in a minimal amount of time.

In general, an optimal control task can be treated either analytically or numerically. In practice, only a small number of problems can be solved analytically, usually small-dimensional problems or special cases; see [20, 21] for two examples. The majority of problems require numerical solutions (see, e.g., [22, 23, 24]), for which it is often advantageous to apply methods from numerical optimisation [25], a field that has developed efficient techniques for dealing with a variety of optimisation problems: continuous or discrete, global or local, constrained or unconstrained, stochastic or deterministic. In this thesis, only deterministic, local and discrete optimisation tasks (with or without constraints) will be presented.

CHAPTER 1: INTRODUCTION

## 1.2  Organisation of this thesis

Chapter 2 introduces the concepts and definitions of group theory that will serve as the foundation for the research in the following chapters. A focus is on Lie groups and their correponding algebras which will play the central role in identifying controllability properties of spin systems in Chapter 3. Here, we study the connection between controllability and symmetry in spin systems of arbitrary topology and with different types of control schemes. After presenting the framework of quantum dynamical control systems, closed quantum systems are characterised in terms of their dynamical Lie algebras. Two examples underpin how inner and outer symmetries impose restrictions on the ability to control systems. The chapter concludes with an investigation of sufficient and necessary conditions for full controllability.

Chapter 4 describes a numerical search for quantum channels that violate a well-known additivity conjecture. The idea of a quantum channel is presented, together with a definition of a channel's capacity. This capacity had been conjectured to be additive or multiplicative, respectively, with implications for classical information theory, until the famous counterexample by Werner and Holevo appeared. Based on their findings, we look for other counterexamples using a gradient-flow on the unitary group to optimise capacities of extremal random unitary channels. In doing so, we focus on channels with similar properties as the Werner-Holevo example. The last part of this chapter is devoted to a summary of the results by Hastings and others who could implicitly show that the additivity conjecture does not hold in the general case.

In Chapter 5 we introduce the optimal control framework and its application to quantum systems, and benchmark an optimal control algorithm known as GRAPE which updates all control variables in a concurrent manner. We specify the algorithmic scheme by analysing the most important components; different procedures for computing the gradient and the update step are presented. Furthermore, we draw a comparison to a Krotov-type sequential-update algorithm and discuss possible hybrid algorithms. Numerical studies with a set of models reveal performance differences between different algorithms and algorithmic components, respectively, and allow for a decision about the best combination of approaches for a given task.

CHAPTER 2

# An introduction to some definitions and concepts of group theory

> ...and you will find someday that, after all, it isn't as horrible.
>
> Richard Feynman

## 2.1 Introduction

This chapter summarises some central concepts of group theory by introducing the definitions and methods needed for the following chapters. A thorough account of group theory can be found in References [26, 27, 28, 29].

## 2.2 Definitions

Given a set $\mathfrak{G} = \{G_1, G_2, \ldots, G_n\}$ of cardinality $n$, and an operation $\circ$ between the elements $G_k, G_l$ of this set:

$$\circ : \mathfrak{G} \times \mathfrak{G} \longrightarrow \mathfrak{G}.$$

The operation is called a multiplication, and $G_k \circ G_l$ is often abbreviated as $G_k G_l$. The algebraic structure $(\mathfrak{G}, \circ)$ is called a *group* if, $\forall\, G_k, G_l \in \mathfrak{G}$,

1. $G_k \circ G_l \in \mathfrak{G}$,
2. the operation is associative: $G_k \circ (G_l \circ G_m) = (G_k \circ G_l) \circ G_m$ ,

## Chapter 2: An Introduction to Some Definitions and Concepts of Group Theory

3. there exists exactly one identity element $E$ for which $E \circ G_k = G_k \circ E = G_k$, and

4. there exists exactly one inverse element $G_k^{-1}$ to each element $G_k$: $G_k G_k^{-1} = E$.

When only the last condition is not fulfilled, $(\mathfrak{G}, \circ)$ is called a *semi-group*. A group is finite, if $n$ is finite; For a group to be continuous it is required that there are infinitely many group elements.

In general, the multiplication of two group elements is non-commutative. Groups with commuting elements, i.e.,

$$G_k \circ G_l = G_l \circ G_k \quad \forall G_k, G_l \in \mathfrak{G},$$

are called *Abelian groups*.

A *homomorphism* between two groups $\mathfrak{G}$ and $\mathfrak{G}'$ is a map that assigns an element $G_k' \in \mathfrak{G}'$ to every element $G_k \in \mathfrak{G}$. Products of $\mathfrak{G}$ are mapped to products of $\mathfrak{G}'$:

$$G_k \to G_k',$$
$$G_l \to G_l',$$
$$G_k \circ G_l \to G_k' \circ G_l'.$$

This map may not be invertible; if it is, one calls it an *isomorphism*.

Two elements $G_k$ and $G_l$ are *conjugate* to each other if there exists a $T \in \mathfrak{G}$ for which $G_l = T G_k T^{-1}$. One writes $G_k \sim G_l$. This relation is

- reflexive: $G_k \sim G_k$,

- symmetric: $G_k \sim G_l \Rightarrow G_l \sim G_k$,

- and transitive: $G_l \sim G_k \wedge G_k \sim G_m \Rightarrow G_l \sim G_m$.

The relation thus divides the group into equivalence classes of conjugate elements.

If the conditions 1-4 are fulfilled for a subset $\mathfrak{H}$ of $\mathfrak{G}$, $\mathfrak{H}$ is called a *subgroup* of $\mathfrak{G}$. A *normal subgroup* $\mathfrak{N}$ is invariant under conjugation with members of the group $\mathfrak{G}$. The *centre* of a group is a subgroup $\mathfrak{Z}$ whose elements commute with all elements of the group:

$$Z \circ G = G \circ Z \quad \forall Z \in \mathfrak{Z}, \; \forall G \in \mathfrak{G}.$$

CHAPTER 2: AN INTRODUCTION TO SOME DEFINITIONS AND CONCEPTS OF GROUP THEORY

Given two groups $\mathfrak{G}_1$ and $\mathfrak{G}_2$, the *direct product* of these groups is the ordered pair

$$\mathfrak{G} := \mathfrak{G}_1 \times \mathfrak{G}_2 = \{(G_1, G_2) \mid G_k \in \mathfrak{G}_k, k = 1,2\},$$

with the operation

$$G \odot G' = (G_1, G_2) \odot (G'_1, G'_2) = (G_1 \circ G'_1, G_2 \circ G'_2).$$

A *simple group* is a nontrivial group that has only two normal subgroups: the trivial group and the group itself.

## 2.3 Matrix representation

A representation $R$ of a group $\mathfrak{G}$ assigns to every element $G_k \in \mathfrak{G}$ a linear map in an $N$-dimensional vector space $V = \mathbb{R}^N$ or $V = \mathbb{C}^N$, i.e., a matrix $R(G_k) \in \text{Mat}_N(\mathbb{C})$, such that the group multiplication corresponds to the matrix multiplication:

$$R(G_k \circ G_l) = R(G_k) R(G_l).$$

We denote the entry in the $k$-th row and $l$-th column of the matrix $R(\cdot)$ as $R(\cdot)_{kl}$. We require that $R(E) = \mathbf{1}_N$ and $R(G_k^{-1}) = R(G_k)^{-1}$. The last requirement makes all matrices non-singular.

If two matrices belong to the same conjugacy class, their traces are equal since the trace is an invariant under a similarity transformation with a matrix $S$:

$$\text{tr}(R(G)) = \text{tr}(S^{-1} R(G) S),$$

where $S$ may be any non-singular matrix.

The outer product between two abstract groups $\mathfrak{G}_1$ and $\mathfrak{G}_2$ corresponds to the matrix tensor product $R(\mathfrak{G}_1) \otimes R(\mathfrak{G}_2)$. The two group representations $R(\mathfrak{G}_1)$ and $R(\mathfrak{G}_2)$ act on the vector spaces $V_1$ and $V_2$, which become $V = V_1 \otimes V_2$ after applying the tensor product.

## 2.4 Lie groups and algebras

### 2.4.1 General

An example for the continuous groups metioned above are *Lie groups* [30]. These groups are *differentiable manifolds*, i.e., group multiplication and inversion are infinitely differentiable with respect to the coordinates in $\mathbb{R}^N$ [31]. They can be thought of as continuous transformations in vector spaces. Hence, many physical problems can be described in terms of Lie groups.

The groups presented in the following are linear, or matrix groups, i.e., groups of linear maps. Common examples for matrix groups are

1. **GL**$(N)$, the full linear group of all non-singular matrices,
2. **SL**$(N)$, the subgroup to **GL**$(N)$ with $\{G \in \mathbf{GL}(N) \,|\, \det(G) = +1\}$,
3. **U**$(N)$, the subgroup to **GL**$(N)$ with $\{G \in \mathbf{GL}(N) \,|\, G^\dagger = G^{-1}\}$,
4. **SU**$(N)$, the subgroup to **U**$(N)$ with $\{G \in \mathbf{U}(N) \,|\, \det(G) = +1\}$,
5. **O**$(N)$, the subgroup to **GL**$(N)$ with $\{G \in \mathbf{GL}(N) \,|\, G^t = G^{-1}\}$, and
6. **SO**$(N)$, the subgroup to **O**$(N)$ with $\{G \in \mathbf{O}(N) \,|\, \det(G) = +1\}$.

The subgroups of the **GL**$(N)$ can be regarded as invariant groups of geometrical entities [30]. For instance, all elements of the group **U**$(N)$ when applied to a vector $v \in \mathbb{C}^N$ leave the length of $v$ invariant. The elements of **U**$(N)$ are thus length-preserving. Another example is the group **SL**$(N)$ whose elements are volume-preserving: The determinant[1] of any matrix $M$ is invariant under the multiplication with an element of **SL**$(N)$.

Every element $U$ of a Lie group can be represented as $U = e^{\varphi x}$, where $\varphi$ is the rotation angle and $x$ the generator of the group element. $x$ can be looked upon as a generalised 'rotation axis'. The exponential map can be defined by the Taylor series

$$e^{\beta x} := \sum_{k=0}^{\infty} \frac{1}{k!} (\beta x)^k.$$

---

[1] $\det(M)$ for $M \in \mathbb{R}^{3 \times 3}$ represents the volume spanned by the three column vectors of $M$.

CHAPTER 2: AN INTRODUCTION TO SOME DEFINITIONS AND CONCEPTS OF GROUP THEORY

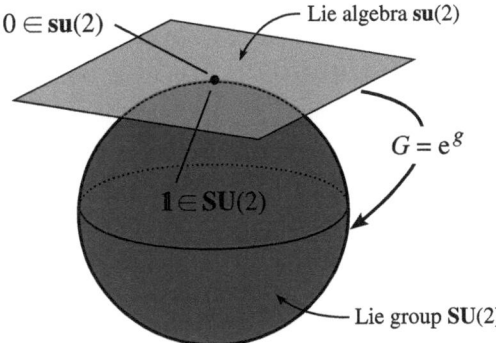

**Figure 2.4.1:** Geometric picture of the Lie group **SU**(2), represented by the surface of the blue sphere, and its corresponding Lie algebra $\mathfrak{su}(2)$, represented by the orange plane. The algebra and the group are connected by the exponential map. Note that both sets are actually 3-dimensional so consider a sphere and a plane embedded in a 4-dimensional space.

The set of all generators $\{x_k\}$ forms a tangent space at the neutral element $\mathbf{1}$ and is called a *Lie algebra* $\mathfrak{su}(N)$, see Figure 2.4.1 and References [30, 31]. This set represents a linear vector space with the scalar product $\mathrm{tr}(a^\dagger b)$ for $a, b \in \mathrm{span}(x_k)$. For composed systems, we find

$$\mathrm{tr}\big((a \otimes b)^\dagger (c \otimes d)\big) = \mathrm{tr}(a^\dagger c)\mathrm{tr}(b^\dagger d).$$

This vector space is closed under commutation, i.e., if $a$ and $b$ are elements of the Lie algebra, so is $[a, b]$. Furthermore, we have

$$[a, b] = -[b, a] \quad \text{and} \tag{2.4.1}$$

$$[a, [b, c]] + [b, [c, a]] + [c, [a, b]] = 0, \tag{2.4.2}$$

where $c$ is again an element of the Lie algebra. The last equation is called *Jacobi identity*.

As a Lie algebra is a linear vector space, it allows for the study of structural properties more easily than on the level of the Lie group. By the surjectivity of the exponential map for compact Lie groups, all properties of the algebra correspond to properties of the group (see Reference [30], pp. 213). This means that results obtained on the level of the algebra can be translated to the group level. We will use this correspondence in Chapter 3 when we

## Chapter 2: An Introduction to Some Definitions and Concepts of Group Theory

investigate controllability of spin systems.

Corresponding to their Lie group names, they are denoted as 1. $\mathfrak{gl}(N)$, 2. $\mathfrak{sl}(N)$, 3. $\mathfrak{u}(N)$, 4. $\mathfrak{su}(N)$, 5. $\mathfrak{o}(N)$, and 6. $\mathfrak{so}(N)$, and contain 1. arbitrary, 2. traceless, 3. skew-Hermitian ($x = -x^{\dagger}$), 4. traceless skew-Hermitian, 5. skew-symmetric ($x = -x^t$), and 6. traceless skew-symmetric matrices.

There exists an isomorphism $f$ between two Lie algebras $a$ and $b$ if all elements of $a$ can be mapped bijectively to all elements of $b$,

$$f((a)_n) = (b)_n \quad \forall\, (a)_n \in a \text{ and } \forall\, (b)_n \in b,$$

and if the same can be done for all commutators:

$$f([(a)_n, (a)_m]) = [(b)_n, (b)_m].$$

As an example, consider the case of $\mathfrak{su}(2)$, in which the commutator can be identified with the vector product of $\mathbf{R}^3$: $(\mathfrak{su}(2), [\cdot, \cdot]) \stackrel{iso}{\cong} (\mathbf{R}^3, \times)$.

Lie algebras can be distiguished from each other by their *structure constants*. Let $\{x_k\}$ be the basis of an algebra. The scalar values $c_{\mu\nu}^{\lambda}$ with

$$[x_\mu, x_\nu] = c_{\mu\nu}^{\lambda} x_\lambda$$

are called *Lie structure constants* [31]. They obey the Equations 2.4.1 and 2.4.2.

Assume $\mathfrak{J}$ is a subspace of the Lie algebra $\mathfrak{g}$: $\mathfrak{J} \subseteq \mathfrak{g}$. If $\mathfrak{J}$ satisfies $[\mathfrak{g}, \mathfrak{J}] \subseteq \mathfrak{J}$, then $\mathfrak{J}$ is called an *ideal* of $\mathfrak{g}$. A non-Abelian Lie algebra, whose only ideals are 0 and itself, is called *simple*. A direct sum of simple Lie algebras is called a *semisimple Lie algebra*. A *simple Lie group* is a connected non-Abelian Lie group without any connected nontrivial normal subgroups.

### 2.4.2 Pauli operator basis

The Pauli matrices

$$\sigma_x := \begin{bmatrix} 0 & 1 \\ 1 & 0 \end{bmatrix} \quad \sigma_y := \begin{bmatrix} 0 & -i \\ i & 0 \end{bmatrix} \quad \sigma_z = \begin{bmatrix} 1 & 0 \\ 0 & -1 \end{bmatrix} \quad (2.4.3)$$

form an orthogonal basis for the Lie algebra $\mathfrak{su}(2)$. One easily verifies that a scalar product exists with $\mathrm{tr}(\sigma_\mu^\dagger \sigma_\nu) = 2\delta_{\mu\nu}$, where $\mu, \nu \in \{x, y, z\}$, and that the commutation relation

$$[\sigma_\mu, \sigma_\nu] = 2i\epsilon_{\mu\nu\xi}\sigma_\xi$$

holds, where $\epsilon_{\mu\nu\xi}$ is the anti-symmetric tensor[2] on three indices and $\mu, \nu, \xi \in \{x, y, z\}$. The Pauli matrices are thus a representation of the generators of $\mathfrak{su}(2)$. In quantum mechanics, the Pauli matrices represent observables describing the spin angular momentum of spin-$\frac{1}{2}$ particles.

A generalisation to Lie algebras $\mathfrak{su}(N)$ and therefore to quantum systems with $n$ spin-$\frac{1}{2}$ particles, where $N = 2^n$ and $n \in \mathbb{N}$, can be obtained in the following way: Taking the tensor product, the unit matrix **1**, and a suitable normalisation, a basis of $\mathfrak{u}(N)$ can be formed by

$$\{\sigma_x^{(1)}, \sigma_y^{(1)}, \sigma_z^{(1)}, \mathbf{1}\} \otimes \{\sigma_x^{(2)}, \sigma_y^{(2)}, \sigma_z^{(2)}, \mathbf{1}\} \otimes \cdots \otimes \{\sigma_x^{(n)}, \sigma_y^{(n)}, \sigma_z^{(n)}, \mathbf{1}\}.$$

Here, $\sigma_\mu^{(n)}$ is the Pauli matrix $\sigma_\mu$ acting on spin $n$. By removing the element $\mathbf{1}_N$, this set becomes a basis of $\mathfrak{su}(N)$.

### 2.4.3 Lie subgroups

Lie groups can have subgroups which can be discrete or continuous. Two examples for subgroups of **SU**(N) will be discussed here.

#### 2.4.3.1 A discrete subgroup of $U(N)$

A trivial example of a discrete subgroup of **SU**(N) is its centre $\mathfrak{z}$. It can be represented by multiples of the identity **1**:

$$\mathfrak{z}(\mathbf{SU}(N)) = \{U = e^{i\phi}\mathbf{1}_N \mid \det(u) = +1\}.$$

The number of elements in $\mathfrak{z}$ is $N$ and the allowed rotation angles are $\phi = 2\pi i k/N$ with $k = 0, 1, \ldots, N-1$.

---

[2] $\epsilon_{\mu\nu\xi} = 0$, except for $\epsilon_{xyz} = \epsilon_{yzx} = \epsilon_{zxy} = 1$, and $\epsilon_{zyx} = \epsilon_{yxz} = \epsilon_{xzy} = -1$

CHAPTER 2: AN INTRODUCTION TO SOME DEFINITIONS AND CONCEPTS OF GROUP THEORY

### 2.4.3.2 A continuous subgroup of $U(N)$

The group of all local operations on $n$ spin-$\frac{1}{2}$ particles,

$$\mathfrak{L} = \mathbf{SU}(2) \otimes \ldots \otimes \mathbf{SU}(2) = \mathbf{SU}(2)^{\otimes n},$$

forms a continuous subgroup of $\mathbf{SU}(N)$. Another example is the group

$$\{U = e^{-i\varphi H} \mid H \in \text{span}\{x1, yz, zz\}\}.$$

Here, $\text{span}\{x1, yz, zz\} = \{\lambda_1 x1 + \lambda_2 yz + \lambda_3 zz \mid \lambda_i \in \mathbb{R}\}$ is the set of linear combinations of the elements $x1, yz,$ and $zz$. The three Pauli terms obey the same commutator relations than the two-dimensional representation of $\mathfrak{su}(2)$; they are hence isomorphic to it.

CHAPTER 3

# Controllability and symmetry in spin systems

> The 'control of nature' is a phrase conceived in arrogance, born of the Neanderthal age of biology and the convenience of man.
>
> Rachel Carson

## 3.1 Introduction

### 3.1.1 Overview

For universal quantum computation on a given physical system, the Hamiltonians have to combine such that any unitary target operation can be performed, irrespective of the initial state of the quantum system. This means the system has to be fully controllable, as has been pointed out in different contexts [32, 33]. Often, the physical system comes in designs with a certain symmetry pattern reflecting the experimental setup. For instance, this is the case in quantum lattices, in quantum networks, or in spin chains serving as quantum wires [34] for distributed quantum computation [35]. Symmetric patterns, however, may have crucial shortcomings since symmetry may prevent full controllability. On the other hand, avoiding symmetry-restricted controllability need not be complicated from an experimental point of view: in Ising spin chains, it soon emerged that polymers $(ABC)_n$ made of three different qubit units $A, B, C$ are fully controllable [36]. Later, irregular $ABAAA\ldots$ systems of just two qubit types $A, B$ [37] and even $(A)_n - B$ chains turned out to be fully controllable as well. In these systems, qubits are meant to be locally controlled by operations that act jointly on all qubits labelled with the same letter and independently from controls on qubits with a different letter. In particular, quantum systems coupled by Heisenberg-$XXX$ type interactions turned out to be fully controllable when the local controls are limited to a small subset of qubits, as with time these actions can then be 'swapped' to neighbouring spins [38, 39, 40, 41, 42, 34, 43].

Such a gradual case-by-case development asks for a more systematic investigation on the quantum side, given the methods of Lie theory to assess controllability in classical systems [44, 45, 38]. Thus, we address both issues with and without symmetry restrictions in qubit systems, where the coupling topology is generalised to any connected graph [46, 47], going beyond linear chains [48] or 'infective' graphs [49].

On a more general scale, it is important to be able to separate questions of existence (e.g.: *Is the system fully controllable? Can it be used for universal Hamiltonian simulation?*) from questions of actual implementation (*How does one have to steer a given experimental setup to implement a target task with highest precision?*). Otherwise, the structure of constructive proofs of existence may translate into highly suboptimal experimental schemes. Hence, here we exploit Lie theoretical methods for a unified framework addressing controllability in a first step, while resorting to quantum control [50] in a second step for actual implementation optimised for the given experimental setup.

### 3.1.2 Organisation and main results

This chapter starts out by introducing quantum dynamical control systems in Section 3.2. Next, we show how symmetries are often easy to see in the coupling graph; as soon as these symmetries have a representation within the dynamical group in question, they preclude full controllabilty, see Sections 3.3 and 3.4. By explicitly identifying the dynamical system algebras in cases of symmetry-constrained controllability, the exact reachable sets take the form of subgroup orbits under the dynamical group generated by the system algebra. This is an important step towards exploring *task controllability* by describing which tasks are feasible on which type of quantum system (Section 3.5). Systems with inner and outer symmetries, including explicit examples, are discussed in Section 3.6.

In turn, the absence of any symmetry is only a necessary condition for full controllability. Since sufficient conditions in the most general case are tedious to come by, we give practical

CHAPTER 3: CONTROLLABILITY AND SYMMETRY IN SPIN SYSTEMS

guidelines for quantum system design in common types of Ising and Heisenberg coupling in Section 3.7. Depending on the coupling type, we will make precise the three design rules sufficient to ensure full controllability in *non-symmetric* (and non-antisymmetric) qubit systems with a connected coupling toplogy:

1. in Ising-$ZZ$ coupled systems *each qubit* has to belong to a *type* that is jointly controllable locally,

2. in Heisenberg-$XXX$ ($XXZ$, $XYZ$) coupled qubit systems *at least one qubit* has to be fully controllable locally,

3. in Heisenberg-$XX$ ($XY$) coupled qubit systems *at least one adjacent qubit pair* has to be fully controllable (by an $\mathfrak{su}(4)$).

We also briefly sketch how to control systems efficiently (Section 3.8), and how to extend notions of controllability for systems with relaxation.

## 3.2 Quantum dynamical control systems

In the following, we address Markovian dynamics of quantum systems, the free evolution of which is governed by a system Hamiltonian $H_0$ and, in the case of open systems, by an additional relaxation term $\Gamma$. Whenever we talk about controllability, we mean *full operator controllability*, thus neglecting more specialised notions like pure-state controllability [51].

In contrast to the system Hamiltonian which can never be 'switched off', the interplay between the quantum system and the experimenter is included by 'switchable' control Hamiltonians $H_m$. They express external manipulations in terms of the quantum system itself, where each control Hamiltonian can be steered in time by control amplitudes $u_m(t)$. With these definitions, the usual equations of motion for controlled quantum dynamics can be brought into a common form, as will be shown next.

As a starting point, consider the Schrödinger equations

$$|\dot\psi(t)\rangle = -i\Big(H_0 + \sum_{m=1}^{M} u_m(t)H_m\Big)\,|\psi(t)\rangle \qquad (3.2.1)$$

$$\dot U(t) = -i\Big(H_0 + \sum_{m=1}^{M} u_m(t)H_m\Big)\,U(t),$$

CHAPTER 3: CONTROLLABILITY AND SYMMETRY IN SPIN SYSTEMS

where the second equation can be regarded as the operator equation to the first one. It governs the evolution of a unitary map of an entire basis set of vectors representing pure states. Using the short-hand notations $H_{tot} := H_0 + \sum_{m=1}^{M} u_m(t) H_m$ and $\mathrm{ad}_H(\cdot) := [H, (\cdot)]$ in the master equation

$$\dot{\rho} = -i[H_{tot}, \rho(t)] - \Gamma(\rho(t)) \equiv -(i\,\mathrm{ad}_{H_{tot}} + \Gamma)\rho(t) \qquad (3.2.2)$$

$$\dot{F}(t) = -(i\,\mathrm{ad}_{H_{tot}} + \Gamma) \circ F(t), \qquad (3.2.3)$$

the second one can again be regarded as the lifted operator equation to the first one: while $\rho \in \mathfrak{her}(N)$, $F$ denotes a quantum map in $GL(N^2)$ as a linear image over all basis states of the Liouville space representing the open system.

All these equations of motion have the form of a standard *bilinear control system* ($\Sigma$) known in classical system and control theory. It reads

$$\dot{X}(t) = \left(A + \sum_{m=1}^{M} u_m(t) B_m\right) X(t), \qquad (3.2.4)$$

with 'state' $X(t) \in \mathbb{C}^N$, drift $A \in \mathrm{Mat}_N(\mathbb{C})$, controls $B_m \in \mathrm{Mat}_N(\mathbb{C})$, and control amplitudes $u_m \in \mathbb{R}$. For simplicity, consider for the moment its linear counterpart with $v \in \mathbb{C}^N$

$$\dot{X}(t) = AX(t) + Bv, \qquad (3.2.5)$$

which is known to be *fully controllable* if it obeys the rank condition (see, e.g., [52])

$$\mathrm{rank}\,[B, AB, A^2 B, \ldots, A^{N-1} B] = N. \qquad (3.2.6)$$

Now lifting the bilinear control system ($\Sigma$) to group manifolds [53, 54] by $X(t) \in GL(N, \mathbb{C})$ under the action of some compact connected Lie group $\mathbf{K}$ with Lie algebra $\mathfrak{k}$ (while keeping $A, B_m \in \mathrm{Mat}_N(\mathbb{C})$), the condition for full controllability turns into its analogue known as the *Lie algebra rank condition* [44, 45, 54]

$$\langle A, B_m \mid m = 1, 2, \ldots, M \rangle_{\mathrm{Lie}} = \mathfrak{k}. \qquad (3.2.7)$$

In this expression, $\langle \cdot \rangle_{\mathrm{Lie}}$ denotes the *Lie closure* obtained by repeatedly taking mutual commutator brackets.

In order to comply with the terminology in Lie theory, we keep the term *Lie dimension* for the real dimension of a Lie algebra, while the *rank of a Lie algebra* is the dimension of its maximal Cartan subalgebra. For instance, $\mathfrak{su}(N)$ has dimension $N^2 - 1$ and rank $N - 1$.

## 3.3 Full controllability and symmetry-restricted controllability in closed quantum systems

In the dynamics of closed quantum systems, the system Hamiltonian $H_0$ is the only drift term, whereas the $H_m$ are again the control Hamiltonians. To fix notations, in systems of $n$ qubits we define $N := 2^n$, so these Hamiltonians $i H_\nu \in \mathfrak{su}(N)$ each generate a one-parameter unitary group of time evolution $\{U_\nu(t) := e^{-itH_\nu} \mid t \in \mathbb{R}^+\} \subset SU(N)$.

Transferring the classical result [45] to the quantum domain [55], the bilinear system of Equation 3.2.1 is *fully operator controllable* if and only if the drift and controls are a generating set of $\mathfrak{su}(N)$:

$$\langle iH_0, iH_m \mid m = 1, 2, \ldots, M\rangle_{\text{Lie}} = \mathfrak{k} = \mathfrak{su}(N). \tag{3.3.1}$$

In fully controllable systems, to every initial state $\rho_0$ the *reachable set* is the entire unitary orbit

$$\mathcal{O}_U(\rho_0) := \{U\rho_0 U^\dagger \mid U \in SU(N)\}.$$

With density operators being Hermitian, this means any final state $\rho(t)$ can be reached from any initial state $\rho_0$ as long as both of them share the same spectrum of eigenvalues.

In contrast, in systems with restricted controllability the Hamiltonians generate a proper subalgebra of the full unitary algebra:

$$\langle iH_0, iH_m \mid m = 1, 2, \ldots, M\rangle_{\text{Lie}} = \mathfrak{k} \subsetneq \mathfrak{su}(N). \tag{3.3.2}$$

### 3.3.1 Algorithm for computing the Lie closure

Suppose we have an $n$-qubit bilinear control system characterised by the drift and control Hamiltonians $\{H_0; H_1, \ldots, H_M\}$. Then the Algorithm 3.1 tabulated above constructively determines a basis of the associated dynamical Lie algebra [56]. Our implementation codes the tensor product basis of Pauli matrices as quaternions simply represented by the Clifford algebra $\mathcal{C}\ell_2(\mathbb{R})$ of quarternary numbers $\{0, 1, 2, 3\}$ plus the Clifford multiplication rules. This allows the calculation of Lie brackets without any matrix operations. For identifying linearly

CHAPTER 3: CONTROLLABILITY AND SYMMETRY IN SPIN SYSTEMS

---
**Algorithm 3.1** Determine the Lie closure for an $n$-qubit system with a given set of drift (or system) and control Hamiltonians. The algorithm is of complexity $\mathcal{O}(256^n)$ for $n$ qubits ($N^2$ rank-revealing QR decompositions with $N = 2^n$).
Start with the inital basis $B_\nu := \{H_0; H_1, \ldots, H_M\}$.
WHILE $M + 1 < \dim \mathfrak{su}(2^n)$
    Perform all Lie brackets $K_i = [H_j, H_k]$ of all elements of the current basis.
    FOR each new $K_i$
        Check linear independence from span $B_\nu$.
        Extend basis by independent new $K_i$: $B_{\nu+1} := \{K_i, H_i\}$.
    ENDFOR
    IF no new $K_i$ found
        Terminate.
    ENDIF
ENDWHILE

---

independent generators, the time consuming step in each iteration is the rank determination by QR decomposition of a sparse coefficient matrix $\mathcal{K} \in \mathrm{Mat}_{4^n}$ collecting all the expansion coefficients to the $K_i$ of Algorithm 3.1 columnwise as vec $(K_i)$ [57]. Our results were cross-checked with GAP 4.4.10 [58].

## 3.3.2 Notation: coupling graphs and interactions

Here, we represent the physical system to the bilinear control system of Equations 3.2.1 and 3.2.4 by a graph $G(V, E)$ see, e.g., Figure 3.3.1. The vertices $V$ of this graph represent the qubits and edges $E$ denote non-vanishing pairwise couplings of Ising or Heisenberg nature. The qubits are taken to be *jointly* affected by local operations that act typewise on all qubits with the same letter and independently from those with a different letter. Sometimes, every qubit type is fully controllable; in other instances, some types are not controlled at all.

More precisely, for $\mu \in \{x, y, z\}$ let

$$\sigma_\mu^{(k)} := \mathbf{1}_2^{\otimes(k-1)} \otimes \sigma_\mu \otimes \mathbf{1}_2^{\otimes(n-k)} \tag{3.3.3}$$

denote the embedded Pauli matrix $\sigma_\mu$ on the $k^{\text{th}}$ qubit of an $n$-qubit system. Then the couplings extend over the edges of the graph $(k, \ell) \in G(V, E)$ and sum all pairwise terms

$$J_{k\ell} \left( \alpha \cdot \sigma_x^{(k)} \sigma_x^{(\ell)} + \beta \cdot \sigma_y^{(k)} \sigma_y^{(\ell)} + \gamma \cdot \sigma_z^{(k)} \sigma_z^{(\ell)} \right). \tag{3.3.4}$$

In the Heisenberg-$XXX$ type, $\alpha = \beta = \gamma \neq 0$, whereas in the $XXZ$ type $\alpha = \beta \neq \gamma$, and in

# CHAPTER 3: CONTROLLABILITY AND SYMMETRY IN SPIN SYSTEMS

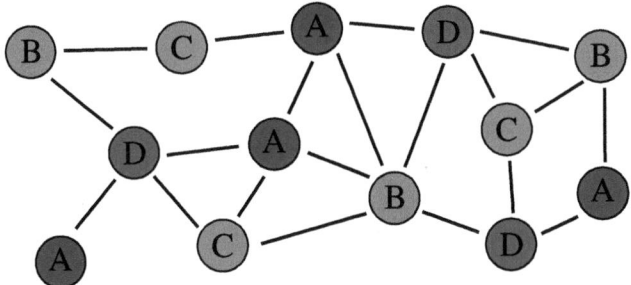

**Figure 3.3.1:** General coupling topology represented by a connected graph. The vertices denote the spin-$\tfrac{1}{2}$ qubits, while the edges represent pairwise couplings of Heisenberg or Ising type. Qubits of the same colour and letter are taken to be affected by joint local unitary operations as in Table 3.1 (or none: see Table 3.2), while qubits of different kind can be controlled independently. For a system to show an outer symmetry brought about by permutations within subsets of qubits of the same type, both the graph as well as the system plus all control Hamiltonians have to remain invariant. In contrast, a system has an outer anti-symmetry, if the edges connected to the nodes of the same type change sign under such a permutation.

the $XYZ$ type $\alpha \neq \beta \neq \gamma \neq \alpha$. In contrast, in the Heisenberg-$XX$ type $\gamma = 0 \neq \alpha = \beta$, while in the $XY$ type $\alpha \neq \beta$. Finally, in an Ising-$ZZ$ interaction $\gamma \neq 0 = \alpha = \beta$.

For a fixed direction $\mu \in \{x,y,z\}$, the local terms are summed over all vertices of the same type in the graph to give the generators $F_\mu := \tfrac{1}{2} \sum_k \sigma_\mu^{(k)}$ of typewise joint local actions.

### 3.3.3 Characterisation by symmetry and antisymmetry

In the following, we will characterise systems of restricted controllability in terms of symmetries. In the present setting, a Hamiltonian quantum system is said to have a symmetry expressed by the skew-Hermitian operator $s \in \mathfrak{su}(N)$, if

$$[s, H_\nu] = 0 \quad \text{for all} \quad \nu \in \{0; 1, 2, \ldots, M\}. \tag{3.3.5}$$

More precisely, we use the term *outer symmetry* if $s$ generates a SWAP operation permuting a subset of qubits of the same type (cp. Figure 3.3.1) such that the coupling graph and all Hamiltonians $\{H_\nu\}$ are left invariant.

Moreover, the coupling Hamiltonian is said to have an *outer anti-symmetry* if there is a permutation $\Pi$ of a subset of vertices $V_\pi \subseteq V$ of the same type such that $\Pi$ leaves the graph

## CHAPTER 3: CONTROLLABILITY AND SYMMETRY IN SPIN SYSTEMS

invariant, while some of the couplings connected to one of the vertices permuted change their respective signs simultaneously:

$$\Pi\, G(V,E) = G(V,E)$$
$$J_{k\ell} = -J_{\Pi(k\ell)} \quad \text{for some} \quad k \in V_\pi \text{ xor } \ell \in V_\pi.$$

For illustration, a simple example (discussed in detail later) can be found in Figure 3.6.2 (a), where an anti-symmetry arises if $J' = -J$.

In contrast, an *inner symmetry* relates to elements $s$ not generating a SWAP operation in the symmetric group of all permutations of qubits in the system; rather than relying on the coupling graph, they are due to the internal structure of drift and control Hamiltonians.

In either case of symmetry, a symmetry operator is an element of the centraliser (or synonymously the commutant)

$$\{H_\nu\}' := \{s \in \mathfrak{su}(N) \mid [s, H_\nu] = 0\ \forall \nu \in \{d; 1, 2, \ldots, M\}\}. \tag{3.3.6}$$

Recall that the centraliser or commutant of a given subset $\mathfrak{m} \subseteq \mathfrak{g}$ with respect to a Lie algebra $\mathfrak{g}$ consists of all elements in $\mathfrak{g}$ that commute with all elements in $\mathfrak{m}$. By Jacobi's identity

$$[[a,b],c] + [[b,c],a] + [[c,a],b] = 0,$$

one gets two properties of the centraliser that are relevant in this context:

1. An element $s$ that commutes with the Hamiltonians $\{iH_\nu\}$ also commutes with their Lie closure $\mathfrak{k}$. For the dynamical Lie algebra $\mathfrak{k}$ we have

$$\mathfrak{k}' := \{s \in \mathfrak{su}(N) \mid [s,k] = 0\ \forall k \in \mathfrak{k}\} \tag{3.3.7}$$

and $\{iH_\nu\}' = \mathfrak{k}'$. Thus, in practice it is convenient to just evaluate the centraliser for a (minimal) generating set $\{iH_\nu\}$ of $\mathfrak{k}$.

2. For a fixed $k \in \mathfrak{k}$, an analogous argument gives

$$[s_1, k] = 0 \quad \text{and} \quad [s_2, k] = 0 \implies [[s_1, s_2], k] = 0, \tag{3.3.8}$$

so the centraliser $\mathfrak{k}'$ forms itself a Lie subalgebra to $\mathfrak{su}(N)$ consisting of all symmetry operators.

CHAPTER 3: CONTROLLABILITY AND SYMMETRY IN SPIN SYSTEMS

---

**Algorithm 3.2** Determine the centraliser to a given set of drift and control Hamiltonians $\{iH_0; H_1, \ldots, H_M\}$. The complexity of the algorithm is $\mathcal{O}(64^n)$ for $n$ qubits, as $4^n$ equations with real coefficients have to be solved by an LU decomposition.

---

FOR each Hamiltonian $H_\nu \in \{H_0; H_1, \ldots, H_M\}$
  determine all solutions to the homogeneous linear eqn.

$$\mathcal{S}_\nu := \{s \in \mathfrak{su}(N) | (\mathbf{1} \otimes H_\nu - H_\nu^t \otimes \mathbf{1}) \operatorname{vec}(s) = 0\}$$

ENDFOR
Obtain the centraliser by intersecting all sets of solutions $\mathfrak{k}' = \bigcap_\nu \mathcal{S}_\nu$.

---

Note that anti-symmetry cannot immediately be detected by evaluating the centraliser as there is no infinitesimal generator in the connected component that would bring about a sign inversion of the coupling term while leaving the local controls invariant. To capture anti-symmetry, we define as the *augmented centraliser* $\mathfrak{k}'_\|$ the centraliser arising after substitution of all coupling constants (but not their anisotropy parameters defined in Equation 3.3.4 for Heisenberg couplings) by their respective moduli

$$(J_{k\ell}, \alpha, \beta, \gamma) \mapsto (|J_{k\ell}|, \alpha, \beta, \gamma).$$

Then, an anti-symmetry in (parts of) the coupling Hamiltonian reveals itself if the original centraliser $\mathfrak{k}'$ is a proper subset of the *augmented centraliser* $\mathfrak{k}'_\|$, so $\mathfrak{k}' \subsetneq \mathfrak{k}'_\|$. In a slight abuse of language, we say a dynamical system has 'no symmetry and no anti-symmetry', if the centraliser is trivial both before and after the above substitution.

Computationally, the centraliser as well as the augmented one are 'exponentially' easier to come by as is evident by comparing the complexity $\mathcal{O}(256^n)$ of Algorithm 3.1 for the Lie closure with the complexity $\mathcal{O}(64^n)$ of Algorithm 3.2 for the centraliser tabulated above. The mere decision whether the centraliser is trivial (without specifying $\mathfrak{k}'$) is of complexity $\mathcal{O}(2^n)$.

Within the centraliser $\mathfrak{k}'$ one may choose a maximally Abelian subalgebra $\mathfrak{a}$ of mutually commuting symmetry operators which allows for a block-diagonal representation in the eigenspaces associated to the eigenvalues $(\lambda_1, \lambda_2, \ldots, \lambda_\ell)$ to $\{a_1, a_2, \ldots, a_\ell\} = \mathfrak{a}$. A convenient set of symmetry operators representing the outer symmetries are the ones generating SWAP transpositions of qubits: they correspond to the $S_2$ symmetry and come with the eigenvalues $+1$ (*gerade*) and $-1$ (*ungerade*). A block-diagonal representation results if all the SWAP transpositions that can be performed independently are taken as one entry each in the $\ell$-tuple $(\lambda_1, \lambda_2, \ldots, \lambda_\ell)$, while all those that have to be performed jointly

# Chapter 3: Controllability and Symmetry in Spin Systems

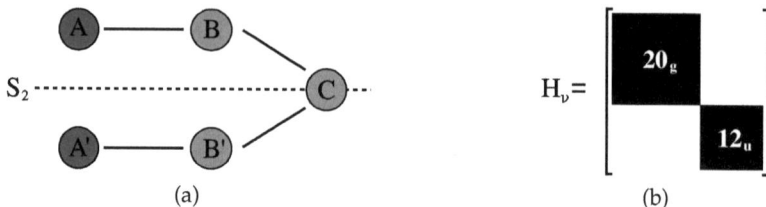

**Figure 3.4.1:** (a) Example 1 with Ising qubit chain of joint $S_2$ symmetry. (b) The drift and control Hamiltonians of Example 1 take block-diagonal form corresponding to the $A_g$ and $A_u$ representation of the $S_2$-symmetry group.

are multiplied together to make one single entry in the tuple. This procedure is illustrated below in Examples 1 and 2.

Observe that in our notation a block-diagonal representation of a Lie algebra $\mathfrak{k} = \mathfrak{su}(N_1) \oplus \mathfrak{su}(N_2)$ generates a group $\mathbf{K} = SU(N_1) \oplus SU(N_2)$ in the sense of a block-diagonal Clebsch-Gordan decomposition (or cartesian product $SU(N_1) \times SU(N_2)$).

## 3.4 Introductory examples with symmetry-restricted controllability

### 3.4.1 Example 1: joint $S_2$ symmetry

First, consider Ising $n$-qubit chains with odd numbers of qubits such as the one in Figure 3.4.1(a), which has an Ising coupling graph $L_5$. It shows a mirror or inversion symmetry $S_2$ (a.k.a. $C_i$) that leaves the coupling graph and thus all Hamiltonians of drifts and controls $\{H_0, H_1, \ldots, H_M\}$ invariant under the joint permutation of qubits $A \leftrightarrow A'$ together with $B \leftrightarrow B'$ in the system of Figure 3.4.1(a) with $C$ in the mirror axis. The joint permutation operator can conveniently be represented as

$$\Pi(L_5) = \begin{bmatrix} 1 & 0 & 0 & 0 \\ 0 & 0 & 1 & 0 \\ 0 & 1 & 0 & 0 \\ 0 & 0 & 0 & 1 \end{bmatrix}_{A,A'} \otimes \begin{bmatrix} 1 & 0 & 0 & 0 \\ 0 & 0 & 1 & 0 \\ 0 & 1 & 0 & 0 \\ 0 & 0 & 0 & 1 \end{bmatrix}_{B,B'} \otimes \mathbf{1}_C \qquad (3.4.1)$$

CHAPTER 3: CONTROLLABILITY AND SYMMETRY IN SPIN SYSTEMS

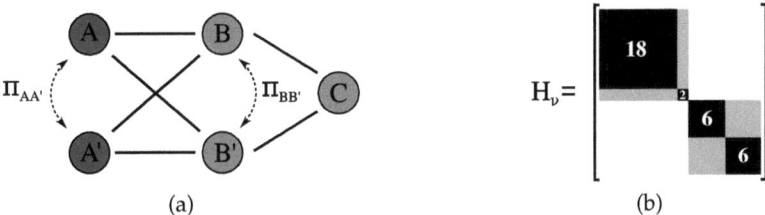

Figure 3.4.2: (a) Example 2 with coupling topology allowing for individually independent permutation symmetries $\Pi_{AA'}$ and $\Pi_{BB'}$, which together reproduce the $S_2$-symmetry of Example 1. (b) The drift and control Hamiltonians of Example 2 take a block-diagonal form corresponding to the *gg*, *uu*, *gu* and *ug* parities of the individual permutations $\Pi_{AA'}$ and $\Pi_{BB'}$.

in a Hilbert space $\mathcal{H}_A \otimes \mathcal{H}_{A'} \otimes \mathcal{H}_B \otimes \mathcal{H}_{B'} \otimes \mathcal{H}_C$ ordered by qubits $A, A', B, B', C$. Figure 3.4.1(b) illustrates how in the $S_2$-symmetry-adapted eigenbasis of $\Pi(L_5)$, all Hamiltonians take the same block-diagonal form with two blocks of parities: *gerade* (eigenvalue $+1$) and *ungerade* (eigenvalue $-1$). There are no further independent constants of motion $s$ in the centraliser to the dynamical system algebra $\mathfrak{k}'_1$ besides the generator of $\Pi(L_5)$. Each block represents a fully controllable logical subsystem with one block being of size $20 \times 20$ and one block of size $12 \times 12$. Note that all the symmetrised Hamiltonians are traceless within each of the two blocks. Thus, the Lie closure gives the Clebsch-Gordan (CG) decomposed dynamical system Lie algebra

$$\mathfrak{k}_1 = \mathfrak{su}(20) \oplus \mathfrak{su}(12) \quad \text{generating} \quad \mathbf{K}_1 = SU(20) \oplus SU(12)$$

with Lie dimension $542 = 399 + 143$.

Example 1 will be generalised to $S_2$-symmetric Ising qubit chains of arbitrary length in Section 3.6.1 below.

## 3.4.2 Example 2: individual permutation symmetry

By introducing further Ising couplings between qubits $A$ and $B'$ and qubits $A'$ and $B$, the $L_5$ system of Example 1 can be turned into the one represented by a non-planar graph in Figure 3.4.2(a): note that now the qubit pairs $A, A'$ and $B, B'$ can be permuted individually. Thus, the block-diagonal representation consists of four blocks corresponding to the parities *gg*, *uu* and *gu*, *ug* with the respective sizes $18 \times 18$, $2 \times 2$ and twice $6 \times 6$, see Figure 3.4.2.

## Chapter 3: Controllability and Symmetry in Spin Systems

The indistinguishable pair $AA'$ (and $BB'$) can be looked upon as a pseudo spin-1 and pseudo spin-0 system, because symmetrisation allows for adding their spin angular momenta in the usual Clebsch-Gordan way. The Lie dimension of each pair is 4, instead of 15 in a fully controllable spin pair; the Lie dimension of the total system is $364 = 323 + 35 + 6$, i.e., the sum of all blockwise Lie dimensions - with two exceptions: (i) the second $6 \times 6$ block is not of full dimension since it does not contain any coupling terms; instead, it only collects elements from two independent $\mathfrak{su}(2)$ subalgebras which arise from the $AA'$ pair and the $C$ spin thus giving a Lie dimension of 6; (ii) the $2 \times 2$ block does not contribute since it reduplicates one of the independent $\mathfrak{su}(2)$ algebras already occuring in the second $6 \times 6$ block, which becomes obvious as the matrix elements in both blocks only occur jointly. Finally, since in all Hamiltonian components all the blocks are independently traceless, we have a CG-decomposed dynamical algebra

$$\mathfrak{k}_2 = \mathfrak{su}(18) \oplus \mathfrak{su}(2) \oplus \mathfrak{su}(6) \oplus \underbrace{\left(\mathfrak{su}(2)_{j=1} \widehat{\oplus} \mathfrak{su}(2)\right)}, \qquad (3.4.2)$$

generating a dynamical group

$$\mathbf{K}_2 = SU(18) \oplus SU(2) \oplus SU(6) \oplus \underbrace{\left(SU(2)_{j=1} \otimes SU(2)\right)}, \qquad (3.4.3)$$

where the index $j = 1$ denotes the spin-1 representation of $\mathfrak{su}(2) \subset \mathfrak{su}(3)$ and the bracket connects the $\mathfrak{su}(2)$ of spin $C$ occuring in two copies (see above).

## 3.5 Task controllability

A set of non-trivial symmetry operators precludes full controllability. With symmetry restrictions in the system, the dynamical Lie algebra $\mathfrak{k}$ is a proper subalgebra of $\mathfrak{su}(N)$:

$$\langle \exp \mathfrak{k} \rangle =: \mathbf{K} \subsetneq SU(N)$$

The reachable sets take the form of *subgroup orbits*

$$\mathcal{O}_K(\rho_0) := \{K\rho_0 K^{-1} | K \in \mathbf{K}\}.$$

Thus, symmetry analysis allows for the specification of the dynamical Lie algebra and for giving selection rules that govern state transfer: An initial quantum state represented by the

# CHAPTER 3: CONTROLLABILITY AND SYMMETRY IN SPIN SYSTEMS

density operator $\rho_0$ can be transferred into a target state $\rho_T$ by a dynamical system with Lie algebra $\mathfrak{k}$ with full fidelity if and only if

$$\rho_T \in \mathcal{O}_{\mathbf{K}}(\rho_0).$$

Note that the situation is easy whenever the states both share the same symmetry as the dynamical system algebra, i.e.,

$$[\rho_0, \mathfrak{k}'] = [\rho_T, \mathfrak{k}'] = 0.$$

Then, $\rho_0$ and $\rho_T$ allow for the same symmetry-adapted block-diagonal decomposition as the system algebra $\mathfrak{k}$, and if for a dynamical group

$$\mathbf{K} = SU(N_1) \oplus SU(N_2) \oplus \cdots \oplus SU(N_\nu)$$

with $N_1 + N_2 + \cdots + N_\nu = N$ also the eigenvalues of $\rho_0$ and $\rho_T$ coincide in each of the $\nu$ blocks, then $\rho_T \in \mathcal{O}_{\mathbf{K}}(\rho_0)$. This is a sufficient condition, not a necessary one.

For illustration, consider the representation of $\rho_0$ and $\rho_T$ in the symmetry-adapted basis of $\mathfrak{k}$ in the easy example of $K \in [SU(N_1) \oplus SU(N_2)]$ with $N_1 + N_2 = N$, where (with $A, \tilde{A}$ and $C, \tilde{C}$ all being Hermitian)

$$\rho_0 := \begin{bmatrix} A & B \\ B^\dagger & C \end{bmatrix}_\mathfrak{k}, \quad \rho_T := \begin{bmatrix} \tilde{A} & \tilde{B} \\ \tilde{B}^\dagger & \tilde{C} \end{bmatrix}_\mathfrak{k}, \quad K := \begin{bmatrix} U & 0 \\ 0 & V \end{bmatrix}_\mathfrak{k}.$$

Then, $K\rho_0 K^\dagger = \rho_T$ holds if and only if simultaneously

$$UAU^\dagger = \tilde{A} \quad \text{and} \quad VCV^\dagger = \tilde{C} \quad \text{and} \quad UBV^\dagger = \tilde{B}.$$

Thus, it is necessary that in the diagonal blocks (here $A, \tilde{A}$ and $C, \tilde{C}$) the eigenvalues coincide blockwise, while in the off-diagonal blocks (here $B, \tilde{B}$) the singular values coincide.

Apart from giving selection rules for state transfers, analysing the dynamical Lie algebra allows for deciding if a specific task is feasible in systems of reduced controllability. A Hamiltonian quantum system characterised by $\{iH_\nu\}$ is called *task controllable* with respect to a target unitary gate $U_G$ if there is at least one Hamiltonian $H_G$ on some branch of the 'logarithm' of $U_G$ so that $U_G = e^{i\phi} \cdot e^{-iH_G}$ (with arbitrary phase $\phi$) and $iH_G \in \mathfrak{k} = \langle iH_\nu \rangle_{\text{Lie}}$ [59]. Whether the Hamiltonian $iH_G$ can be generated by a system with dynamical Lie algebra $\mathfrak{k}$, can be tested by simply evaluating a matrix rank: arrange all the matrices $\{k_1, k_2, \ldots, k_r\}$

CHAPTER 3: CONTROLLABILITY AND SYMMETRY IN SPIN SYSTEMS

spanning $\mathfrak{k}$ as column vectors collected in a matrix $K := \begin{bmatrix} \operatorname{vec}(k_1), \operatorname{vec}(k_2), \ldots, \operatorname{vec}(k_r) \end{bmatrix}$. Then $iH_G$ can be generated in $\mathfrak{k}$ if rank $(K)$ = rank $\begin{bmatrix} K, \operatorname{vec}(iH_G), \operatorname{vec}(\mathbf{1}) \end{bmatrix}$.

## 3.6 Discussion of inner and outer symmetries

Here, we further study dynamical Lie algebras of systems with outer and inner symmetries. Recall that outer symmetries permute equivalent qubits in the coupling graph, while inner symmetries reflect constants of motion that are due to the Hamiltonians themselves rather than due to permuting among qubits.

This section is structured as follows. In 3.6.1, we discuss Ising-$ZZ$ coupled systems with local control on all qubits as examples, where outer symmetries are the only pertinent ones. In 3.6.2, the focus is on Heisenberg-$XXX$ coupled systems with local control on just a single qubit as examples of inner symmetries. In 3.6.3, we investigate qubit chains with minimalistic local controls on one or two qubits at the controlled end (called *head part* henceforth) and Heisenberg-$XY$ type couplings throughout the uncontrolled part. These systems are of interest as breaking their inner symmetries explains controls that are necessary for an exponential growth of the reachable state space.

### 3.6.1 Systems with outer symmetry

Outer symmetries directly relate to permutations of vertices in the network's representation by a coupling graph. Thus, they are easy to see just as in the introductory examples of Figures 3.4.1 and 3.4.2. Here, we extend them to a larger set of instances given in Table 3.1. For linear chains, the only applicable symmetry is the joint $C_i$ mirror operation permuting the first and the second half of the chain in the sense of the symmetric group $S_2$. This gives rise to a block-diagonal irreducible representation of the dynamical algebra with just two blocks: they are associated to the $+1$ and the $-1$ eigenspace of the permutation, or the *gerade* and *ungerade* subspace of $C_i$ (see above). This explains instances (a) through (f) plus (m) of Table 3.1. Closing chains to cycles does not change the situation except for allowing for unimportant phase factors as in (i) compared to (j). Note that in the four-membered ring (h) the isotropic Heisenberg coupling reduces the Lie dimension, whereas in the five-membered cycle of (j) it does not. When a loop is introduced as in case (l), the mirror symmetry is broken and the system becomes fully controllable. Case (k) has been treated separately in all detail

**Table 3.1:** Dynamical system algebras to qubit systems with joint local controls on each type (letter) and various couplings

| | System | Coupling Types | Lie Dim. | Block Sizes in irred. rep. | Lie Dims. blockwise | System Lie Algebra | Trace = 0 blockwise |
|---|---|---|---|---|---|---|---|
| (a) | $A - A$ | ZZ, XX, XY | 9 | 3, 1 | 9, 1 | $\mathfrak{s}(\mathfrak{u}(3) \oplus \mathfrak{u}(1))$ | no |
| (b) | $A - A$ | XXX | 4 | 3, 1 | 4, 1 | $\mathfrak{s}(\mathfrak{u}(2)_{j=1} \oplus \mathfrak{u}(1))$ | no |
| (c) | $A - B$ | -all- | 15 | 4 | 15 | $\mathfrak{su}(4)$ | yes |
| (d) | $A \mid B$ | -none- | 6 | 4 | 6 | $\mathfrak{su}(2) \widehat{\oplus} \mathfrak{su}(2)$ | yes |
| (e) | $A - B - A$ | -all- | 38 | 6, 2 | 35, 3 | $\mathfrak{su}(6) \oplus \mathfrak{su}(2)$ | yes |
| (f) | $A - B - B - A$ | -all- | 135 | 10, 6 | 100, 36 | $\mathfrak{s}(\mathfrak{u}(10) \oplus \mathfrak{u}(6))$ | no |
| (g) | $\underline{A - B - B - A}$ | ZZ, XX, XY | 135 | 10, 6 | 100, 36 | $\mathfrak{s}(\mathfrak{u}(10) \oplus \mathfrak{u}(6))$ | no |
| (h) | $\underline{A - B - B - A}$ | XXX | 115 | 10, 6 | 100, 16 | $\mathfrak{s}(\mathfrak{u}(10) \oplus \mathfrak{u}(4) \subset \mathfrak{u}(6))^{\sharp}$ | no |
| (i) | $A - B - C - B - A$ | -all- | 542 | 20, 12 | 399, 143 | $\mathfrak{su}(20) \oplus \mathfrak{su}(12)$ | yes |
| (j) | $\underline{A - B - C - B - A}$ | -all- | 543 | 20, 12 | 400, 144 | $\mathfrak{s}(\mathfrak{u}(20) \oplus \mathfrak{u}(12))$ | no |
| (k) | $\underline{A - B - C - B - A}$ | -all- | 364 | 18, 2, 6, 6 | 323, (3), 35, 6 | see lengthy Equation (3.4.2) | yes |
| (l) | $\underline{A - B - C - B - A}$ | -all- | 1023 | 32 | 1023 | $\mathfrak{su}(32)$ | yes |
| (m) | $A - B - C - C - B - A$ | -all- | 2079 | 36, 28 | 1296, 784 | $\mathfrak{s}(\mathfrak{u}(36) \oplus \mathfrak{u}(28))$ | no |

($\sharp$) Here $\mathfrak{u}(4) \subset \mathfrak{u}(6)$ denotes an irreducible representation of $\mathfrak{u}(4)$ embedded in $\mathfrak{u}(6)$.

CHAPTER 3: CONTROLLABILITY AND SYMMETRY IN SPIN SYSTEMS

as Example 2 above.

#### 3.6.1.1 Arbitrary $n$-qubit chains with reflection symmetry

For an Ising $n$-qubit chain $L_n$ with central mirror symmetry $S_2$, the findings of the introductory Example 1 generalise, thus providing a common formula that covers the results in Table 3.1(a), (e), (f), (i), and (m). Let $p := \lfloor n/2 \rfloor$ define the number of qubit pairs. Then the joint permutation $j \leftrightarrow (n-j+1)$ for all qubits $j = 1, \ldots, p$ may be represented in the basis of Equation 3.4.1 as

$$\Pi(L_n) = \begin{bmatrix} 1 & 0 & 0 & 0 \\ 0 & 0 & 1 & 0 \\ 0 & 1 & 0 & 0 \\ 0 & 0 & 0 & 1 \end{bmatrix}^{\otimes p} \otimes \mathbf{1}_{1+(n \bmod 2)}, \tag{3.6.1}$$

where the last factor is the number 1 for $n$ even and the unit operator $\mathbf{1}_2$ on the central qubit for $n$ odd. The generator of $\Pi(L_n)$, up to a phase $\phi$ so that $e^{i\phi}\Pi(L_n) \in SU(N)$, is the only non-trivial element in the centraliser $\mathfrak{k}'(L_n)$ of the dynamical Lie algebra $\mathfrak{k}(L_n)$ to an $n$-qubit chain with mirror symmetry. Thus, in the $S_2$-symmetry adapted bases the dimensions $d_g$ and $d_u$ of the respective *gerade* and *ungerade* subspaces are determined as follows: each two-qubit SWAP in Equation 3.6.1 contributes the eigenvalues $+1$ (3-fold degenerate) and $-1$ (non-degenerate). For a product of $p$ pairs, the respective dimensions $d_g$ and $d_u$ of the gerade and ungerade subspaces collect alternating terms of a binomial distribution

$$d_g(n) := 2^{(n \bmod 2)} \sum_{k=0}^{\lfloor p/2 \rfloor} \binom{p}{2k} 3^{p-2k} \tag{3.6.2}$$

$$d_u(n) := 2^{(n \bmod 2)} \sum_{k=0}^{\lfloor p/2 \rfloor - 1} \binom{p}{2k+1} 3^{p-2k-1}. \tag{3.6.3}$$

The binomial distribution inherently ensures $d_g + d_u = 2^n$ by the row sum in Pascal's triangle.

We may generalise Example 1 to

**Proposition 1.** *In general Ising- or Heisenberg-coupled n-qubit chains $L_n$ with central reflection*

## CHAPTER 3: CONTROLLABILITY AND SYMMETRY IN SPIN SYSTEMS

symmetry $S_2$ and hence pairwise locally controllable qubits, the dynamical Lie algebras are

$$\mathfrak{k}(L_n) = \begin{cases} \mathfrak{su}(d_g(n)) \oplus \mathfrak{su}(d_u(n)) & \text{for } n \text{ odd} \\ \mathfrak{s}[\mathfrak{u}(d_g(n)) \oplus \mathfrak{u}(d_u(n))] & \text{for } n \text{ even} \end{cases}$$

with $d_g(n)$ and $d_u(n)$ as in Equations 3.6.2 and 3.6.3.

*Proof.* First, we prove the dimension formulae of Equations 3.6.2 and 3.6.3: Symmetry-adapted bases of the SWAP operator of Equation 3.6.1 show block-diagonal form with a core of size $3 \times 3$ and $1 \times 1$:

$$\begin{bmatrix} \begin{array}{|ccc|} \hline \cdot & \cdot & \cdot \\ \cdot & 3_g & \cdot \\ \cdot & \cdot & \cdot \\ \hline \end{array} & \\ & \boxed{1_u} \end{bmatrix}^{\otimes p} \otimes \mathbf{1}_{1+n(\text{mod}2)}. \tag{3.6.4}$$

For the dimensions, the $p^{\text{th}}$ tensor power yields a binomial distribution $(1_u + 3_g)^p = \sum_{\nu=0}^{p} \binom{p}{\nu}(1_u^\nu \cdot 3_g^{p-\nu})$. The formulae then follow by summing the terms $\nu = 2k$ of parity gerade for $d_g$ and the terms $\nu = 2k+1$ of parity ungerade for $d_u$.

Second, we prove the two cases observing that Equation 3.6.4 implies the recursions

$$d_g(n+2) = 3 \cdot d_g(n) + d_u(n)$$
$$d_u(n+2) = 3 \cdot d_u(n) + d_g(n),$$

as does the above binomial distribution.

(1) For $n$ odd, the induction $n \mapsto n+2$ may be based on the $A - B - A$ system of Table 3.1(e) with the dynamical system algebra

$$\mathfrak{k}(L_3) = \mathfrak{su}(6) \oplus \mathfrak{su}(2) \equiv \mathfrak{su}(d_g(n)) \oplus \mathfrak{su}(d_u(n)).$$

In the new eigenbasis of $\Pi(L_{n+2})$, the coupling term connecting the new qubit pair $(\Omega, \Omega')$ to the terminal qubit pair $(A, A')$ of the old chain $L_n$ reads as

$$((z1)_{\Omega,\Omega'} \otimes (z1)_{A,A'} + (1z)_{\Omega,\Omega'} \otimes (1z)_{A,A'}) \otimes \mathbf{1}_2^{\otimes(n-2)}$$

CHAPTER 3: CONTROLLABILITY AND SYMMETRY IN SPIN SYSTEMS

and can be chosen diagonal. It has non-vanishing elements in both new blocks, the one of dimension $d_g(n+2)$ as well as $d_u(n+2)$. Note that 3 Pauli-basis elements that formerly were among $d_g(n)$ (respectively $d_u(n)$) now have elements in both $d_g(n+2)$ an $d_u(n+2)$ and form a subalgebra $\mathfrak{su}(2)$ among themselves. They can be coupled to the new 3 Pauli-basis elements on $(\Omega, \Omega')$. Hence, to each element that formerly was among $d_g(n)$ there are 3 new ones in $d_g(n+2)$ that arise from coupling to $(\Omega, \Omega')$ via the $d_g(n+2)$ part of the coupling term plus local actions on the $d_g(n+2)$ part of $(\Omega, \Omega')$. The ones formerly in $d_g(n)$ and now in $d_u(n+2)$ correspond to the untransformed elements. They add to the new ones in $d_u(n+2)$, which analogously arise via coupling: they are 3 times as many as in the former $d_u(n)$ part thus giving $d_u(n+2) = 3 \cdot d_u(n) + d_g(n)$. The ones formerly in $d_u(n)$ and now in $d_g(n+2)$ count again as untransformed to make a total of $d_g(n+2) = 3 \cdot d_g(n) + d_u(n)$, so the recursion formulae are exactly matched.

(2) For $n$ even, the induction $n \mapsto n+2$ is based on the $A - B - B - A$ system of Table 3.1(f) with the dynamical system algebra

$$\mathfrak{k}(L_4) = \mathfrak{s}\big(\mathfrak{u}(10) \oplus \mathfrak{u}(6)\big) \equiv \mathfrak{s}\big[\mathfrak{u}(d_g(n)) \oplus \mathfrak{u}(d_u(n))\big],$$

where the same arguments apply. The additional phase degree of freedom compared to (1) is due to the coupling between the innermost qubit pair, hence arising only in chains with even $n$. □

In view of engineering quantum systems, we will close the subsequent paragraphs by a summarising guideline. They will be made more rigorous below. The first one is:

**Design Rule 1.** *For an Ising-ZZ coupled n-qubit system to be fully controllable, it suffices that*

1. *each qubit belongs to a type that is jointly operator controllable locally (as in Figure 3.3.1),*

2. *the coupling topology forms a connected graph, and*

3. *drift and control Hamiltonians share no outer symmetries or anti-symmetries and $\mathfrak{k}'$ is trivial.*

## 3.6.2 Systems with inner symmetry

In contradiction to permutation-type outer symmetries, inner symmetries are not immediately obvious from the coupling topology and its representation as a graph. They rather

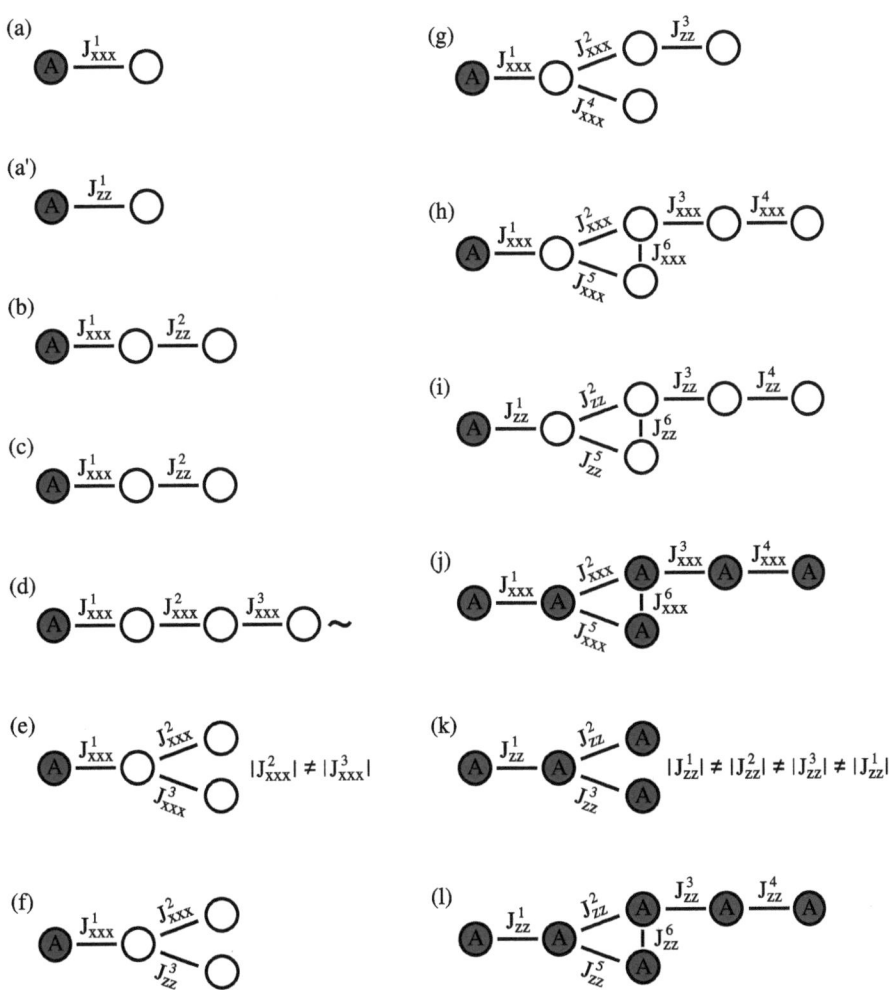

**Figure 3.6.1:** Coupled qubit systems with subsets of uncontrolled qubits that are interacting via a connected coupling topology of Heisenberg and Ising type. Their dynamical Lie algebras are listed in Table 3.2. Note that the graphs of (e), (h), (k) and (l) are all 'non-infective' [49], yet the systems are fully controllable. Under collective controls, Ising interactions in non-symmetric systems (k) and (l) give full controllability, while isotropic Heisenberg interactions (j) do not. Full controllability in (j) requires non-isotropic $XXZ$-type couplings.

**Table 3.2:** Dynamic system algebras to qubit systems with subsets of uncontrolled qubits (cf. Figure 3.6.1)

| System shown in | Coupling types non-zero $J^i$ | Lie dim. | Block sizes in irred. rep. | Lie dims. blockwise | System Lie algebra | Trace = 0 blockwise |
|---|---|---|---|---|---|---|
| Figure 3.6.1(a) | $J_{xxx}^1$ | 15 | 4 | 15 | $\mathfrak{su}(4)$ | yes |
| Figure 3.6.1(a') | $J_{zz}^1$ | 6 | 2 | 3,3 | $\mathfrak{su}(2)\oplus\mathfrak{su}(2)$ | yes |
| Figure 3.6.1(b) | $J_{xxx}^1, J_{zz}^2$ | 30 | 4,4 | 15,15 | $\mathfrak{su}(4)\oplus\mathfrak{su}(4)$ | yes |
| Figure 3.6.1(c) | $J_{xxx}^1, J_{zz}^2$ | 30 | 4,4 | 15,15 | $\mathfrak{su}(4)\oplus\mathfrak{su}(4)$ | yes |
| Figure 3.6.1(d) | $J_{xxx}^i$ | 255 | 16 | 255 | $\mathfrak{su}(16)$ | yes |
| Figure 3.6.1(e) | $J^1_{xxx} J^2_{xxx} \neq J^3_{xxx}$ | 255 | 16 | 255 | $\mathfrak{su}(16)$ | yes |
| Figure 3.6.1(f) | $J_{xxx}^{1,2}, J_{zz}^3$ | 126 | 8,8 | 63,63 | $\mathfrak{su}(8)\oplus\mathfrak{su}(8)$ | yes |
| Figure 3.6.1(g) | $J_{xxx}^{1,2,3}, J_{zz}^4$ | 510 | 16,16 | 255,255 | $\mathfrak{su}(16)\oplus\mathfrak{su}(16)$ | yes |
| Figure 3.6.1(h) | $J_{xxx}^i$ | 4095 | 64 | 4095 | $\mathfrak{su}(64)$ | yes |
| Figure 3.6.1(i) | $J_{zz}^i$ | 7 | $2_{32}^{(\dagger)}$ | $4_{32}^{(\dagger)}$ | $\mathfrak{s}(\mathfrak{u}(2)\oplus\mathfrak{u}(2))_{16}^{(\dagger)}$ | no |
| Figure 3.6.1(j) | $J_{xxx}^i$ equal or unequal | 4 | $7_1,5_5,3_9,1_5$ | — | $\mathfrak{s}(\mathfrak{u}(2)\oplus\mathfrak{u}(1))$ | no |
| Figure 3.6.1(k) | $J_{zz}^i$ all unequal | 255 | 16 | 255 | $\mathfrak{su}(16)$ | yes |
| Figure 3.6.1(l) | $J_{zz}^i$ all equal | 4095 | 64 | 4095 | $\mathfrak{su}(64)$ | yes |

$^{(\dagger)}$ Indices $4_k$ and $\mathfrak{su}(4)_k$ denote $k$ identical block sizes and $k$ identical block-diagonal copies of the same algebra, respectively.

CHAPTER 3: CONTROLLABILITY AND SYMMETRY IN SPIN SYSTEMS

arise as Hamiltonian symmetries due to a lack of control combined with particular coupling types, thus reflecting constants of motion. For instance, if a qubit's polarisation is conserved, one is faced with what is called a *passive qubit*. Note this terminology is in line with well-established definitions in magnetic resonance spectroscopy [60]. In Figure 3.6.1 and Table 3.2, any system that is not fully controllable shows passive qubits $p_i$ attached via Ising couplings, except the special case (j). Every individual passive qubit $p_i$ comes with a symmetry operator $\sigma_z^{(p_i)}$ in the centraliser. In the instances of Table 3.2(a'), (b), (c), (f), and (g), there is just one single such passive qubit and therefore the block-diagonal irreducible representation of the dynamic Lie algebra is made of two equivalent blocks associated with the $+1$ and the $-1$ eigenspace of $\sigma_z^{(p_i)}$. If several qubits are passive, then the block-diagonal irreducible representation is structured by the eigenvalues $\lambda_{pi} = \pm 1$ of the independent $\sigma_z^{(p_i)}$ operators sorted as tuples $(\lambda_{p1}, \lambda_{p2}, \ldots, \lambda_{pr})$. With the number of such tuples being $2^r$ for $r$ passive qubits, the irreducible representation takes block-diagonal form with $2^r$ components. In the instance (i) of Figure 3.6.1 and Table 3.2, there are five such passive qubits and hence $2^5 = 32$ blocks of $4 \times 4$ size. Due to the further Ising couplings, cases (a') and (i) differ by one degree of freedom arising from non-traceless blocks reflecting an unimportant phase factor.

In contrast, the case of collective controls on isotropically Heisenberg-coupled qubits is an extreme example, with the drift Hamiltonian commuting with the control Hamiltonians being shown as instance (j) in Figure 3.6.1 and Table 3.2: here, the collective local controls are trivially invariant under permutation of qubits. Since the isotropic Heisenberg coupling Hamiltonian $H_{xxx}$ is the generator of the SWAP operation, the coupling Hamiltonian as a drift term commutes with the collective controls thus leaving a dynamic algebra $\mathfrak{k}$ of Lie dimension four: it is generated by the three local joint qubit rotations along the $x, y, z$ axes plus the isotropic coupling term which introduces an immaterial phase. An equivalent example with two indistinguishable qubits already occured in Table 3.1(b). More precisely, the irreducible representation of the dynamic Lie algebra of the six-qubit system in Table 3.2(j) is merely a block-diagonal decomposition of $\mathfrak{su}(2)$, where the drift term introduces an immaterial phase.

For full controllability in Ising-coupled systems, the symmetry may be broken either by unequal coupling constants (instance (k) in Figure 3.6.1 and Table 3.2 or by a non-symmetric coupling topology, as in instance (l). Finally, note that (j) can be made fully controllable if the isotropic Heisenberg interaction is replaced by a non-isotropic one like $XXZ$. Otherwise, $H_{XXX} \in \mathfrak{k}'$ since it commutes with all joint controls.

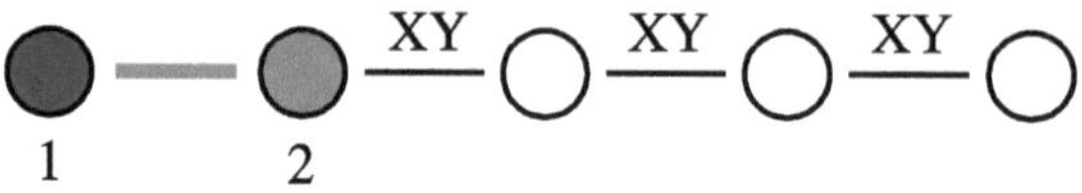

# CHAPTER 3: CONTROLLABILITY AND SYMMETRY IN SPIN SYSTEMS

**Figure 3.6.2:** Qubit chains with minimalistic controls in the two-qubit head and an uncontrolled remainder with Heisenberg-$XX$ and $-XY$ type coupling drifts as discussed in Table 3.7.1 referring to recent literature [43, 61, 62].

**Design Rule 2.** *For a generic[1] Heisenberg-XXX (XXZ, XYZ) coupled n-qubit system to be fully controllable, it suffices that*

1. *one single qubit is fully controllable locally,*

2. *the coupling topology forms a connected graph, and*

3. *drift and control Hamiltonians share no outer symmetries or anti-symmetries and $\mathfrak{k}'$ is trivial.*

These conditions are less restrictive than requiring a coupling topology with an 'infecting graph' [49].

## 3.6.3 Qubit chains with minimalistic controls

Here, we compare $XX$-coupled qubit chains with different types and degrees of control. For this purpose, in Table 3.3, a variety of setups with different parameters $\delta_i$ and $B_i$ in

$$H_d := \tfrac{1}{2} \sum_{k=2}^{n-1} (1+\delta) X_k X_{k+1} + (1-\delta) Y_k Y_{k+1} + \sum_{i=2}^{n} B_i Z_i$$

are collected. These systems are similar to the ones recently studied in [61, 62], but we confine ourselves to pure qubit systems and do not aim for simulating fermionic or bosonic systems on such qubit systems.

Recall from Equation 3.3.3 the $\sigma_z^{(i)}$ as the embedded Pauli matrix. Then the symmetry operators $\mathfrak{k}'$ to the respective dynamic Lie algebras $\mathfrak{k}$ in Table 3.3(a) through (f) all comprise the operator

$$P_z := \prod_{i=1}^{n} \sigma_z^{(i)}$$

---

[1] Heisenberg-$XYZ$ type coupling interaction is called generic if in the tuple $(\alpha, \beta, \gamma)$ associated to the coupling term $J_{k\ell} \, (\alpha \cdot \sigma_x^{(k)} \sigma_x^{(\ell)} + \beta \cdot \sigma_y^{(k)} \sigma_y^{(\ell)} + \gamma \cdot \sigma_z^{(k)} \sigma_z^{(\ell)})$ there are no pairs just differing in sign, since we have not analysed particular instances like $\beta = -\alpha = \pm \gamma$ etc.

**Table 3.3:** Dynamic system algebras to qubit chains with minimalistic controls and various types of couplings (cf. Figure 3.6.2).

| | Controls | Drift Pars $B_j$ | Drift Pars $\delta$ | Lie Dim. | Block Sizes in irred. rep. | Lie Dims. blockwise | System Lie Algebra | Trace = 0 blockwise |
|---|---|---|---|---|---|---|---|---|
| (a) | $XX_{12}$ | 0 | 0 | 10 | 1,5,10,10,5,1 | 0,10,10,10,10,0 | $\mathfrak{so}(5)$ | yes |
| (b) | $XY_{12}$ | 0 | 0.3 | 20 | 16, 16 | 20, 20 | $\mathfrak{so}(5) \oplus \mathfrak{so}(5)$ | yes |
| (c) | $Z_1$ (plus $XY_1$ in $H_0$) | 1 | 0 | 25 | 1,5,10,10,5,1 | 1,25,25,25,25,1 | $\mathfrak{s}(\mathfrak{u}(5) \oplus \mathfrak{u}(1))$ | no |
| (d) | $Z_1$ (plus $XY_1$ in $H_0$) | 1 | 0.3 | 45 | 16, 16 | 45, 45 | $\mathfrak{so}(10)$ | yes |
| (e) | $XX_{12}$ | 1 | 0 | 11 | 1,5,10,10,5,1 | 1,11,11,11,11,1 | $\mathfrak{s}(\mathfrak{o}(5) \oplus \mathfrak{u}(1))$ | no |
| (f) | $XY_{12}$ | 1 | 0.3 | 45 | 16, 16 | 45, 45 | $\mathfrak{so}(10)$ | yes |
| (g) | $Z_1, X_1, XX_{12}, ZZ_{12}^{(b)}$ | 0 | 0 | $1023^{(b)}$ | 32 | $1023^{(b)}$ | $\mathfrak{su}(2^5)$ | yes |
| (g') | $Z_1, X_1, XXX_{12}^{(b)}$ | 0 | 0 | $1023^{(b)}$ | 32 | $1023^{(b)}$ | $\mathfrak{su}(2^5)$ | yes |
| (h) | $Z_1, X_1, XY_{12}, ZZ_{12}^{(b)}$ | 1 | 0.3 | $1023^{(b)}$ | 32 | $1023^{(b)}$ | $\mathfrak{su}(2^5)$ | yes |
| (h') | $Z_1, X_1, XYZ_{12}^{(b)}$ | 1 | 0.3 | $1023^{(b)}$ | 32 | $1023^{(b)}$ | $\mathfrak{su}(2^5)$ | yes |

[b] Without the controls $ZZ_{12}$ (separate or in $XXX_{12}$) the Lie dimensions reduce to 55 in either of the cases (g,g') and (h,h').

## Chapter 3: Controllability and Symmetry in Spin Systems

whose eigenvalues $\pm 1$ already contribute a separation into two block-diagonal components. Moreover, instances (a), (c), and (e) are characterised by the additional symmetry operator (here with $n = 5$)

$$F_z := \tfrac{1}{2} \sum_{i=1}^{n} \sigma_z^{(i)},$$

whose eigenvalues are conserved quantities. The block-diagonal irreducible representations of the dynamic system algebras consist of six blocks associated to these eigenvalues $p \in \tfrac{1}{2}\{5, 3, 1, -1, -3, -5\}$.

Relaxing the Heisenberg-$XX$ coupling to $XY$ (hereby setting $\delta = 0.3$) breaks this symmetry. So instead of $F_z$, in the instances (b), (d), (f) one is just left with the symmetry operator $P_z$, whose eigenvalues $\pm 1$ lead to an irreducible representation with two blocks of equal size.

In order to discuss cases (a) and (b) explicitly, consider the analogous three-qubit system, which in the case of $XX$ couplings ($\delta = 0$) takes block-diagonal form of sizes $1, 3, 3, 1$, whereas under $XY$ ($\delta \neq 0$) coupling the block sizes are $4, 4$. In the eigenbases of $F_z$, one finds (with the short-hands $c_\delta, d_\delta = \delta c, \delta d$) two useful equivalent representations

$$c \cdot H_c + d \cdot H_d = \begin{bmatrix} 0 & 0 & -ic_\delta & id_\delta & 0 & 0 & 0 & 0 \\ 0 & 0 & d & 0 & 0 & 0 & 0 & 0 \\ -ic_\delta & -d & 0 & c & 0 & 0 & 0 & 0 \\ id_\delta & 0 & -c & 0 & 0 & 0 & 0 & 0 \\ 0 & 0 & 0 & 0 & 0 & c & 0 & id_\delta \\ 0 & 0 & 0 & 0 & -c & 0 & d & -ic_\delta \\ 0 & 0 & 0 & 0 & 0 & -d & 0 & 0 \\ 0 & 0 & 0 & 0 & id_\delta & -ic_\delta & 0 & 0 \end{bmatrix}$$

$$\simeq \begin{bmatrix} 0 & 0 & 0 & 0 & -id_\delta & ic_\delta & 0 & 0 \\ 0 & 0 & d & 0 & 0 & 0 & 0 & 0 \\ 0 & -d & 0 & c & 0 & 0 & 0 & ic_\delta \\ 0 & 0 & -c & 0 & 0 & 0 & 0 & -id_\delta \\ -id_\delta & 0 & 0 & 0 & 0 & c & 0 & 0 \\ ic_\delta & 0 & 0 & 0 & -c & 0 & -d & 0 \\ 0 & 0 & 0 & 0 & 0 & -d & 0 & 0 \\ 0 & 0 & ic_\delta & -id_\delta & 0 & 0 & 0 & 0 \end{bmatrix}.$$

Thus, for $\delta = 0$ one gets two block-diagonal copies of a joint algebra $\mathfrak{so}(3)$. For $\delta \neq 0$ the matrix representations remain skew-symmetric in the real parts, while their imaginary parts are symmetric. In the latter representation, the real part and the imaginary part constitute a $\mathfrak{k}$-$\mathfrak{p}$ decomposition of the non-compact real form $\mathfrak{so}^*(4)$ to $\mathfrak{so}(4)$ (see, e.g., [63] p. 343) with Lie dimension 6, thus duplicating the Lie dimension of $\mathfrak{so}(3)$ upon departing from $\delta = 0$ to $\delta \neq 0$. In the analogous case of five qubits, see Table 3.3(b), the decomposition is no longer as elementary as before, but for $\delta = 0$ one finds again two block-diagonal copies of a joint $\mathfrak{so}(5)$. For $\delta \neq 0$, the block-diagonal part consists of the zero-quantum interactions of the type

$$H_{ZQ} := \sigma_x^{(j)} \otimes \sigma_x^{(k)} + \sigma_y^{(j)} \otimes \sigma_y^{(k)}$$

whereas the terms outside the block-diagonal correspond to double-quantum interactions of the type

$$H_{DQ} := \sigma_x^{(j)} \otimes \sigma_x^{(k)} - \sigma_y^{(j)} \otimes \sigma_y^{(k)}.$$

In total, one obtains again a duplication of the degrees of freedom to arrive at a dynamic Lie algebra isomorphic to $\mathfrak{so}(5) \widehat{\oplus} \mathfrak{so}(5)$ with overall Lie dimension 20.

In conclusion, the findings of this section can be summarised by the following practical guideline:

**Design Rule 3.** *For a Heisenberg-XX (XY) coupled n-qubit system to be fully controllable, it suffices that*

1. *one adjacent qubit pair is fully controllable as $\mathfrak{su}(4)$,*

2. *the coupling topology forms a connected graph, and*

3. *drift and control Hamiltonians share no outer symmetries or anti-symmetries and $\mathfrak{k}'$ is trivial.*

The design rules 1 through 3 are made more rigorous in the following paragraph.

## 3.7 Absence of symmetry versus full controllability

Ultimately, the question is: under which conditions does the absence of any symmetry imply full controllability? In the special case of pure-state controllability, this interrelation was analysed in [47]. In the generalised context of full operator controllability, the issue was raised in [64], among others, following the lines of [65], however, without a full answer.

CHAPTER 3: CONTROLLABILITY AND SYMMETRY IN SPIN SYSTEMS

Here, we focus on quantum systems where the drift Hamiltonian is comprised of Ising- or Heisenberg-type couplings in a topology that can take the form of any connected graph. We note the following.

### 3.7.1 Absence of symmetry implies (semi-)simplicity

**Lemma 1.** *Let $\mathfrak{k} \subseteq \mathfrak{su}(N)$ be a matrix Lie subalgebra to the compact simple Lie algebra of special unitaries $\mathfrak{su}(N)$. If the centraliser $\mathfrak{k}'$ of $\mathfrak{k}$ in $\mathfrak{su}(N)$ is trivial, then*

*(1) $\mathfrak{k}$ is given in an irreducible representation;*

*(2) $\mathfrak{k}$ is simple or semi-simple.*

*Proof.* (1) The unitary representation of the corresponding matrix Lie group $\mathbf{K} \subseteq SU(N)$ ensures full reducibility invoking the Schur-Weyl theorem. As there is no invariant subspace $\mathfrak{k}'$ other than the trivial ones, the representations of $\mathbf{K}$ and $\mathfrak{k}$ are irreducible.

(2) Since $\mathfrak{k}$ is by construction a Lie subalgebra to the compact Lie algebra $\mathfrak{su}(N)$, $\mathfrak{k}$ is compact itself. By compactness it has a decomposition into its centre and a semi-simple part $\mathfrak{k} = \mathfrak{z}_\mathfrak{k} \oplus \mathfrak{ss}$ (see, e.g., [66] Corollary IV.4.25). As the centre $\mathfrak{z}_\mathfrak{k} = \mathfrak{k}' \cap \mathfrak{k}$ is trivial and $\mathfrak{k}$ is traceless, $\mathfrak{k}$ itself can only be semi-simple or simple. □

### 3.7.2 Conditions for simplicity

**Lemma 2.** *Let the Lie closure $\mathfrak{k}$ of a set of drift and control Hamiltonians $\{iH_\nu\}$ be a compact Lie algebra with trivial centraliser $\mathfrak{k}'$ in $\mathfrak{su}(N)$. If the coupling topology to the drift term $H_0$ takes the form of a connected graph extending over the entire system, then $\mathfrak{k}$ is a simple Lie subalgebra of $\mathfrak{su}(N)$.*

*Proof.* For an $n$-qubit drift Hamiltonian with a coupling topology of a graph that is connected, there exists no representation by a single Kronecker sum. (Rather, it is a linear combination of Kronecker sums). As every semi-simple Lie algebra allows for a representation as a single Kronecker sum, the dynamic Lie algebra $\mathfrak{k}$ can only be simple. □

CHAPTER 3: CONTROLLABILITY AND SYMMETRY IN SPIN SYSTEMS

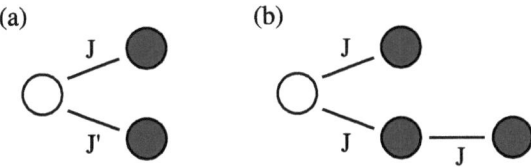

**Figure 3.7.1:** At branching points in the coupling graphs, the (anti-)symmetry between qubits jointly controlled by local operations (coloured nodes) can be broken (a) by different coupling constants $|J| \neq |J'|$ or (b) by different topological continuation as described in the text.

#### 3.7.2.1 Easiest example

Consider an Ising-coupled two-qubit system with individual local controllability, so the Lie closure to $\{iH_v\} \setminus iH_0$ is $\mathfrak{su}(2) \oplus \mathfrak{su}(2)$, which is semi-simple (and isomorphic to $\mathfrak{so}(4)$) and has just a trivial centraliser. Upon including the Ising coupling, the Lie closure of the full $\{iH_v\}$ turns into $\mathfrak{su}(4)$, which is simple.

Therefore, in systems with local controllability a lack of symmetry gives a trivial centraliser, which in turn implies irreducibility. Together with compactness it entails (at least) semi-simplicity, while a connected topology of (appropriate) couplings on top finally ensures simplicity.

### 3.7.3 Sufficient conditions for full controllability

Moreover, the premise of a connected coupling topology together with the conditions specified as Design Rules 1 to 3 even imply full controllability, as summarised in the previous section. To see this, some principal implications have to be established as lemmas.

As a basis, we use the following well-known result:

**Proposition 2.** *For a spin-$\frac{1}{2}$ network to be fully controllable, it suffices that every qubit is fully controllable locally and the coupling topology takes the form of a connected graph, while the coupling may be of Ising-type [46] or Heisenberg-type [47].*

Now, systems with partially collective local controls can be shown to be fully controllable, if the partial symmetry between local controls is broken by the coupling topology.

**Lemma 3.** *As illustrated in the setting of Figure 3.7.1, the symmetry between qubits controlled jointly by local operations can be broken*

# CHAPTER 3: CONTROLLABILITY AND SYMMETRY IN SPIN SYSTEMS

(1) by different coupling constants $|J| \neq |J'|$ or

(2) by a different coupling-topological continuation.

*Proof.* We prove the lemma in three scenarios: (A) for Ising coupled systems with partially joint local controls, (B) for Heisenberg-XXX coupled systems with partially uncontrolled qubits, and (C) for Heisenberg-XX coupled systems also with partially uncontrolled qubits.

## A: Ising coupling

(1) We consider the branching point of Figure 3.7.1(a) as an $A - B - A'$ Ising chain. We define the coupling $H_{zz} := J(zz1) + J'(1zz)$ and the $A$-controls $H_{Ax} := u(x11) + u'(11x)$ on the Hilbert space $\mathcal{H}_A \otimes \mathcal{H}_B \otimes \mathcal{H}_{A'}$. Then the single commutator with $iH_{zz}$

$$\operatorname{ad}_{iH_{zz}}(iH_{Ax}) \equiv [iH_{zz}, iH_{Ax}]$$
$$= -i\{uJ(yz1) + u'J'(1zy)\}$$

is linearly independent of the triple commutator

$$\operatorname{ad}^3_{iH_{zz}}(iH_{Ax}) = i\{uJ^3(yz1) + u'J'^3(1zy)\}$$

unless $\frac{u}{u'}\frac{J}{J'} = \frac{u}{u'}(\frac{J}{J'})^3$, i.e. $(\frac{J}{J'})^2 = 1$. Likewise, $H_{Ax} := u(x11) + u'(11x)$ is linearly independent of

$$\operatorname{ad}^2_{iH_{zz}}(iH_{Ax}) = -i\{uJ^2(x11) + u'J'^2(11x)\}$$

unless again $(\frac{J}{J'})^2 = 1$: such an (anti-)symmetry is, however, excluded by the premise of trivial centralisers $\mathfrak{k}'$ and $\mathfrak{k}'_{\|}$. Once the local controls become independent, Proposition 2 applies. Hence, in view of the Lie dimensions being the rank of the Lie closure seen as a vector space, the symmetry between $A$ and $A'$ can not only be broken by independently switchable controls $u, u'$, but also by different coupling constants $|J| \neq |J'|$ to give

$$\dim \langle iH_{zz}, iH_{Ax} \rangle_{\text{Lie}} = 6$$

instead of 3 in case $|J| = |J'|$. In the same manner, one may consider the extended system with independent $x$ and $y$ controls jointly on $A, A'$. Including independent $x$ and $y$ controls on $B$, one finds a fully controllable three-qubit system in case $|J| \neq |J'|$

$$\dim \langle iH_{zz}, iH_{Ax}, iH_{Ay}, iH_{Bx}, iH_{By} \rangle_{\text{Lie}} = 63.$$

CHAPTER 3: CONTROLLABILITY AND SYMMETRY IN SPIN SYSTEMS

As long as no qubit is without some local control, the same arguments hold for other coupling types of Heisenberg type, where it extends to $(J_{k\ell}, \alpha_{xx}, \beta_{yy}, \gamma_{zz})$ in order to avoid anti-symmetry.

(2) Next, take the setting of Figure 3.7.1 (b) as an $A - B - A' - A''$ Ising chain. Now, with the uniform coupling term $H_{zz} := J\{zz11 + 1zz1 + 11zz\}$ and the joint $A$-controls $H_{Ax} := u(x111 + 11x1 + 111x)$, one finds

$$\mathrm{ad}^2_{iH_{zz}}(iH_{Ax}) = -iuJ^2\{(x111) + 2(11x1) + 2(1zxz) + (111x)\},$$

which, when compared to $H_{Ax}$, shows that the local controls on qubits $A$ and $A'$ are linearly independent. Then, include the $B$-control $H_{Bx} := v(1x11)$ and compare the expression

$$\mathrm{ad}_{iH_{zz}} \circ \mathrm{ad}^2_{iH_{Bx}} \circ \mathrm{ad}_{iH_{zz}}(iH_{Ax}) = iuv^2J^2\{(x111) + (11x1) + (1zxz)\}$$

to the one above to see how the local controls on qubits $A$ and $A''$ become linearly independent, too. Thus, we have reduced the problem to satisfy the preconditions for Proposition 2 which finally makes the entire system fully controllable:

$$\dim \langle iH_{zz}, iH_{Ax}, iH_{Ay}, iH_{Bx}, iH_{By}\rangle_{\mathrm{Lie}} = 255.$$

**B: Heisenberg-$XXX$ coupling**

1. We take the branching point of Figure 3.7.1(a) as an $O - A - O'$ Heisenberg-$XXX$ chain with

$$H_{xxx} := J(xx1 + yy1 + zz1) + J'(1xx + 1yy + 1zz),$$

where the $O$ qubits are uncontrolled, while the $A$ qubit has full local control via $H_{Ax} = u(1x1)$ and $H_{Ay} = v(1y1)$. Then, the double commutator with $iH_{xxx}$ contains the local terms

$$\mathrm{ad}^2_{iH_{xxx}}(iH_{Ax}) = -2iu\{(J^2 + J'^2)(1x1) - J^2(x11) - J'^2(11x)\},$$

which are linearly independent of $H_{Ax}$ and also introduce independent local controls on $O$ and $O'$ when combined with

$$\mathrm{ad}^4_{iH_{xxx}}(iH_{Ax}) = +2iu\{(4J^4 + 10J^2J'^2 + 4J'^4)(1x1)$$
$$- (4J^4 + 5J^2J'^2)(x11) - (4J'^4 + 5J^2J'^2)(11x)\},$$

CHAPTER 3: CONTROLLABILITY AND SYMMETRY IN SPIN SYSTEMS

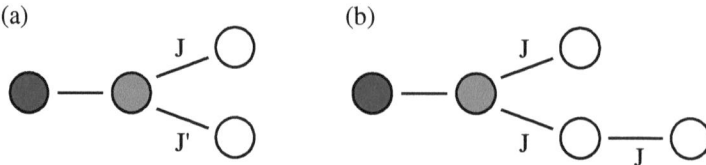

**Figure 3.7.2:** At branching points in the coupling graphs, also the symmetry between uncontrolled qubits (white nodes) can be broken (a) by different coupling constants $|J| \neq |J'|$ or (b) by a different topological continuation as described in the text..

unless $|J| = |J'|$, which is excluded by the premise of trivial centralisers $\mathfrak{k}'$ and $\mathfrak{k}'_{\|}$. Once all the local controls are independent, Proposition 2 applies and the system is fully controllable.

2. Take Figure 3.7.1(b) as an $O - A - O' - O''$ Heisenberg-XXX chain with

$$H_{xxx} := J(xx11 + yy11 + zz11 + 1xx1 + 1yy1 + 1zz1 + 11xx + 11yy + 11zz),$$

where the $O$ qubits are again uncontrolled, while the $A$ qubit has full local control via $H_{Ax} = u(1x11)$ and $H_{Ay} = v(1y11)$. Then, one arrives at

$$\mathrm{ad}^2_{iH_{xxx}}(iH_{Ax}) = -2iuJ^2\{2(1x11) - (x111) - (11x1)\},$$

to give joint local controls on $O$ and $O'$ independent of $H_{Ax}$. Again, the controls on $O$ and $O'$ can then be made independent when combined with

$$\mathrm{ad}^4_{iH_{xxx}}(iH_{Ax}) = +iuJ^4\{40(1x11) - 18(x111) - 28(11x1) + 6(111x)\},$$

and finally the control on $O''$ becomes also independent via $\mathrm{ad}^6_{iH_{xxx}}(iH_{Ax})$ to invoke Proposition 2 for full controllability.

**C: Heisenberg-XX coupling**

1. Consider the topology of Figure 3.7.2(a), where the shaded nodes $A - B$ represent a fully controllable $\mathfrak{su}(4)$ subsystem $XX$-coupled to the uncontrolled qubits $O, O'$ (white nodes) via $H_{XX} := J(1xx1 + 1yy1) + J'(1x1x + 1y1y)$. Then, by

$$\mathrm{ad}^2_{i(1x11)} \circ \mathrm{ad}^2_{i(xx11+yy11)} \circ \mathrm{ad}^2_{iH_{XX}}(i(zz11)) = -4i\{J^2(1zz1) + J'^2(1z1z)\}$$

42

CHAPTER 3: CONTROLLABILITY AND SYMMETRY IN SPIN SYSTEMS

one can supply the Ising terms on $B-O$ and $B-O'$ independently for $|J| \neq |J'|$ to link the problem to the case B(1) solved above. We find

$$\mathrm{ad}_{iJ^2(1zz1)+iJ'^2(1z1z)} \circ \mathrm{ad}_{iH_{XX}}(iH_{Bx}) = iu\,\{J^3(11x1) + J'^3(111x)\}$$

to establish the preconditions for Proposition 2.

2. Likewise, in the system of Figure 3.7.2(b) with the couplings

$$H_{XX} := J(1xx11 + 1yy11) + J(1x1x1 + 1y1y1) + J(111xx + 111yy),$$

one obtains the supplementary Ising terms

$$F_1(i(zz111)) = 2i\,J^2\{(1zz11) + (1z1z1)\}$$

by

$$F_1 := \mathrm{ad}_{i(yy111)} \circ \mathrm{ad}_{i(1y111)} \circ \mathrm{ad}_{i(z1111)} \circ \mathrm{ad}_{i(1x111)} \circ \mathrm{ad}_{i(xx111+yy111)} \circ \mathrm{ad}^2_{iH_{XX}},$$

and moreover, setting

$$F_2 := \mathrm{ad}_{i(1y111)} \circ \mathrm{ad}_{i(1z111)} \circ \mathrm{ad}_{i(1x111)} \circ \mathrm{ad}^2_{iH_{XX}},$$

also the more remote Ising terms

$$F_2 \circ F_1(i(zz111)) = 4i\,J^4\{(1z11z) - (1zz11) - 2(1z1z1)\},$$

and via $F_3 := \mathrm{ad}^2_{iH_{XX}}$ finally

$$F_3 \circ F_2 \circ F_1(i(zz111)) = 8i\,J^6\{(111zz) - 3(11zz1) + (1zz11) + (11z1z) \\ + 5(1z1z1) - 5(1z11z)\}$$

Once they are all made linearly independent, the problem is reduced to the case B(2) treated above.

$\square$

**Theorem 1.** Let $\mathfrak{k} = \langle iH_0, iH_1, \ldots, iH_M\rangle_{\mathrm{Lie}} \subseteq \mathfrak{su}(N)$ be simple with the centraliser $\mathfrak{k}'$ and the augmented centraliser $\mathfrak{k}'_{\|}$ both trivial. Let $H_0$ be of a connected coupling topology in each of the

CHAPTER 3: CONTROLLABILITY AND SYMMETRY IN SPIN SYSTEMS

*coupling types invoked below.* Then, any one of the following additional conditions ensures full controllability:

(1) the system is coupled by Ising-$ZZ$ interactions and each qubit belongs to a type that is jointly operator controllable locally;

(2) the system is coupled by generic Heisenberg-$XXX$ (or $XXZ$, $XYZ$) interactions and there is at least one qubit that is fully controllable locally;

(3) the system is coupled by generic Heisenberg-$XX$ (or $XY$) interactions and there is at least one adjacent qubit pair that is fully controllable in the sense of a $\mathfrak{su}(4)$.

*Proof.* The three conditions are proven separately:

(1) The instance of a locally controllable system coupled via a connected graph of Ising-$ZZ$ interactions was proven to be fully controllable in [46, 47]. The situation does not change for typewise joint local controls, because any permutation symmetry among these joint local controls must be broken by the coupling term to fulfill the premise of a trivial centraliser $\mathfrak{k}'$. Hence, Lemma 3 applies to give the assertion.

(2) The Heisenberg-$XXX$ couplings swap a single fully controllable qubit through the connected coupling network. The local controls are independent as long as no symmetries are introduced so that Lemma 3 applies.

(3) By the same token, the Heisenberg-$XX$ couplings perform a successive iSWAP of a fully controllable qubit pair through the coupling network. The $ZZ$ parts missing after the iSWAP can be corrected for in a second step, since the qubit pair shifted is fully controllable by premise.

□

## 3.7.4 Necessary conditions for full controllability

Having given engineering rules sufficient for full controllability, one would like to proceed a step further: what are the the necessary conditions to fill the gap between lack of symmetry and full controllability in systems of $n$-qubit systems coupled in a connected topology of Ising- or Heisenberg-type interactions? A full treatment will be given in [67], where, based on complete lists of irreducible simple subalgebras of $\mathfrak{su}(N)$, convenient algorithmic schemes will be devised boiling down to solving systems of homogeneous linear equations to identify $\mathfrak{k}$ as $\mathfrak{su}(N)$, thus filtering it from all other potential candidates.

## 3.8 Efficient controllability

So far, we have exploited the power of Lie theory for addressing controllability as an abstract decision problem. For the experimenter, however, systems have to be controlled efficiently. Since there are simple rules of thumb for designing efficient quantum systems, we add them as a final complementary guideline without going into further detail here.

**Design Rule 4.** *Fully controllable quantum systems can be made efficient by ensuring that*

1. *the coupling graph has a small diameter d,*

2. *the couplings are large compared to the fastest relaxation-rate constant, or more precisely, the smallest coupling $J^*$ necessary to maintain connectedness of the coupling graph fulfills $\frac{1}{d}|J^*_{\min}| \gg T_R^{-1}$ (with $T_R$ as the relaxation-rate constant),*

3. *the drift Hamiltonian $H_0$ has well separated eigenvalues,*

4. *the number of separately addressable qubits is not orders of magnitude lower than the total number of qubits.*

## 3.9 Conclusions

In this chapter, we treated controllability in a unified Lie-algebraic framework incorporating constraints by symmetry for closed systems. In particular, the dynamic system Lie algebra allows for the specification of the reachability sets of closed systems explicitly. Our results show that avoiding symmetries can be advantageous wherever they are not explicitly desired, e.g., in order to exploit them for decoherence-protected subspaces or in code spaces for error correction.

In quantum systems with symmetry, the feasible tasks in quantum simulation or quantum gate synthesis can be made precise, thus giving valuable guidelines for quantum system

CHAPTER 3: CONTROLLABILITY AND SYMMETRY IN SPIN SYSTEMS

design matched to solve a given problem. We provided design rules that ensure full controllability in systems with Ising-$ZZ$, Heisenberg-$XX$ ($XY$) and Heisenberg-$XXX$ ($XXZ, XYZ$) type couplings with limited local access.

CHAPTER 4

# Numerical studies on the additivity of quantum channel capacities

> Success is often achieved by those who don't know that failure is inevitable.
>
> — Coco Chanel

## 4.1 Introduction

Can entanglement increase the amount of classical information sent over a quantum channel? This question is strongly related to another one: Is the capacity of a quantum channel additive? Based on a known counterexample by Holevo and Werner [68], we investigate quantum channels for their additivity properties by optimising their capacity measure using gradient flow techniques. Of particular interest are random extremal channels with unitary matrices as their Kraus operators.

Section 4.2 defines quantum channels and gives some examples of standard channels including their representation on the Bloch sphere. The additivity conjecture for the capacities of these channels is presented in Section 4.3, followed by the famous counterexample by Werner and Holevo in Section 4.4. The chapter concludes with a description of our numerical studies and their findings. We briefly outline further analytical work that succeeded our studies and disproved the additivity conjecture for channels in very large dimensions.

We thank Michael Wolf for initiating this project and for his collaboration in the course of this project.

CHAPTER 4: NUMERICAL STUDIES ON THE ADDITIVITY OF QUANTUM CHANNEL
CAPACITIES

## 4.2 Quantum channels

### 4.2.1 Definition

In quantum information theory, a quantum channel is a device for transmitting an input state to an output state. This makes it the basic object of study in this field. Physically, a quantum channel often describes decoherence as introducing noise into a system by entanglement with the environment; mathematically, a quantum channel is defined to be a completely positive trace-preserving (CPTP) map between matrix algebras [69].

Let $M_n$ denote the algebra of complex $n \times n$ matrices. A linear map $\Phi : M_n \to M_m$ is called *completely positive* (CP) if $\Phi \times \mathbf{1}_k : M_n \otimes M_k \to M_m \otimes M_k$ is positivity preserving for every $k \geq 1$, where $\mathbf{1}_k$ is the identity map on $M_k$. Any map $\Phi$ is CP if and only if it can be written in the *Kraus representation*:

$$\Phi(X) := \sum_{i=1}^{L} K_i X K_i^\dagger.$$

Here, $X \in M_n$, and the $m \times n$ matrices $K_1, \ldots, K_L$ are called *Kraus operators* of $\Phi$. The minimum number of Kraus operators is called the *Kraus rank*. The Kraus representation is not unique, but if $K_1, \ldots, K_L$ and $\tilde{K}_1, \ldots, \tilde{K}_{\tilde{L}}$ are two different Kraus representations for the same map with $L \leq \tilde{L}$, then there exists a $L \times \tilde{L}$ matrix $A = (a_{ij})$ such that

$$K_i = \sum_{j=1}^{\tilde{L}} a_{ij} \tilde{K}_j, \quad AA^\dagger = \mathbf{1}_L.$$

Regarding a CP map as an operator acting on quantum states, we note the following: If $\rho$ is a quantum state, i.e., a positive semidefinite operator with trace one, then $\Phi(\rho)$ must also be a quantum state. The CP condition ensures that $\phi(\rho) \geq 0$, but since the probabilities must be preserved, the map is also required to be trace-preserving (TP). For a map that is CP and TP, the Kraus operators must satisfy

$$\sum_{i}^{L} K_i^\dagger K_i = \mathbf{1}_n.$$

If $L = 1$, the matrix $K_1$ is thus unitary and the channel is called *pure*. Time evolution is an example of a pure quantum channel.

For a detailed physical interpretation of quantum channels, see Chapter 8.2 in [2].

## 4.2.2 The Bloch sphere representation

For a single qubit, a quantum channel can be visualised by its action on the Bloch sphere. Since the Bloch representation of a single qubit state is

$$\rho = \frac{\mathbb{1}_2 + \mathbf{r}\sigma}{2} = \frac{1}{2}\begin{bmatrix} 1+r_z & r_x - ir_y \\ r_x + ir_y & 1 - r_z \end{bmatrix},$$

where $\mathbf{r}$ is a real vector with the three components $r_x, r_y, r_z$, any quantum channel can be regarded as the affine map

$$\mathbf{r} \xrightarrow{\Phi} \mathbf{r}' = M\mathbf{r} + \mathbf{c}.$$

Here, $M$ is a $3 \times 3$ real matrix and $\mathbf{c}$ is a constant vector. As explained in Chapter 8.3.2 of [2], this equation describes a deformation of the Bloch sphere.

## 4.2.3 Examples

We briefly introduce some important quantum channels on a single qubit:

The *bit flip channel* flips the qubit state from $|0\rangle$ to $|1\rangle$ with probability $w$. The Kraus operators for this channel are

$$K_1 = \sqrt{w}\,\mathbb{1}_2 = \sqrt{w}\begin{bmatrix} 1 & 0 \\ 0 & 1 \end{bmatrix} \quad \text{and} \quad K_2 = \sqrt{1-w}\,\sigma_x = \sqrt{1-w}\begin{bmatrix} 0 & 1 \\ 1 & 0 \end{bmatrix}.$$

The action of this channel on the Bloch sphere is illustrated in Figure 4.2.1.

Analogously, the *phase flip channel* changes the phase of the qubit with probability $w$:

$$K_1 = \sqrt{w}\,\mathbb{1}_2 = \sqrt{w}\begin{bmatrix} 1 & 0 \\ 0 & 1 \end{bmatrix} \quad \text{and} \quad K_2 = \sqrt{1-w}\,\sigma_z = \sqrt{1-w}\begin{bmatrix} 1 & 0 \\ 0 & -1 \end{bmatrix}.$$

Figure 4.2.2 shows how this channel affects the Bloch sphere.

The *depolarising channel* describes a noisy quantum process: The state of a qubit is depolarised, i.e., replaced with a completely mixed state $\mathbb{1}_2/2$ with probability $w$. The channel

# Chapter 4: Numerical Studies on the Additivity of Quantum Channel Capacities

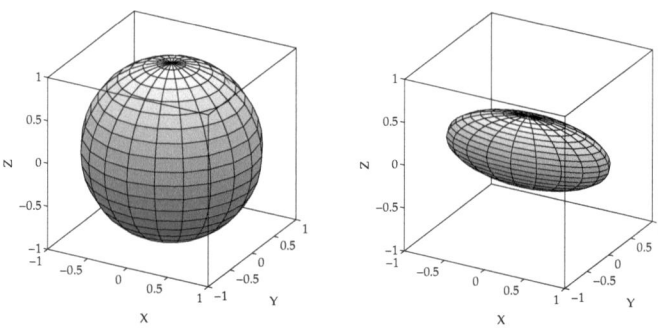

**Figure 4.2.1:** Bloch sphere representation of the bit flip channel with $w = 0.3$. The left sphere represents all pure states, the sphere on the right represents the same set of states after the channel acted on them. Note that the $y$-$z$ plane is uniformly contracted by $1 - 2w$.

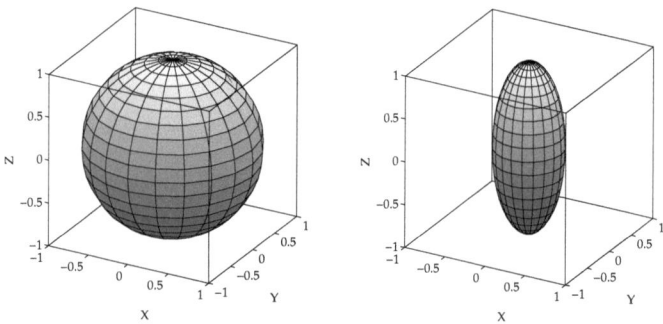

**Figure 4.2.2:** The action of the phase flip channel on the Bloch sphere, for $w = 0.3$. The $x$-$y$ plane is uniformly contracted by $1 - 2w$.

can be represented as

$$\Phi(\rho) = (1-w)\rho + \frac{w}{3}(\sigma_x \rho \sigma_x + \sigma_y \rho \sigma_y + \sigma_z \rho \sigma_z),$$

which shows that the state $\rho$ is left unchanged with probability $1 - w$, and the operators $\sigma_x$, $\sigma_y$, and $\sigma_z$ are applied each with probability $w/3$. The Kraus operators for the depolarising channel are

$$K_1 = \sqrt{1 - 3w/4}\, \mathbf{1}_2 = \sqrt{1 - 3w/4} \begin{bmatrix} 1 & 0 \\ 0 & 1 \end{bmatrix},$$

# CHAPTER 4: NUMERICAL STUDIES ON THE ADDITIVITY OF QUANTUM CHANNEL CAPACITIES

$$K_2 = \sqrt{w}\,\sigma_x/2 = \sqrt{1-w}\begin{bmatrix} 0 & \frac{1}{2} \\ \frac{1}{2} & 0 \end{bmatrix},$$

$$K_3 = \sqrt{w}\,\sigma_y/2 = \sqrt{1-w}\begin{bmatrix} 0 & -\frac{i}{2} \\ \frac{i}{2} & 0 \end{bmatrix},$$

$$K_4 = \sqrt{w}\,\sigma_z/2 = \sqrt{1-w}\begin{bmatrix} \frac{1}{2} & 0 \\ 0 & -\frac{1}{2} \end{bmatrix}.$$

A Bloch sphere representation of the depolarising channel is depicted in Figure 4.2.3.

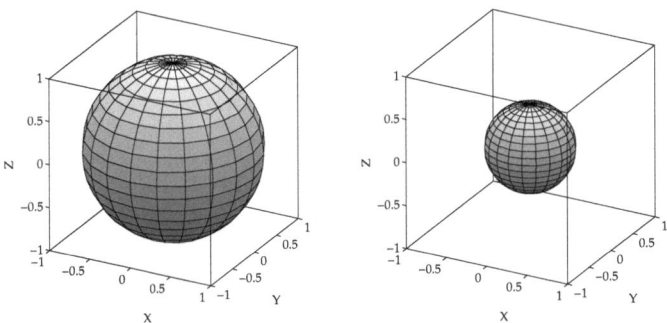

**Figure 4.2.3:** Bloch sphere representation of the depolarising channel with $w = 0.5$. The channel contracts the entire sphere in a uniform way, depending on $w$.

## 4.3 The additivity conjecture

In general, the capacity of a channel is the maximal transmission rate for one use of the channel. This allows one to define different capacities depending on the conditions under which the channel is used. For quantum channels transmitting classical information, a useful representation of the capacity is the highest purity of the channel output. One may then ask: is this quantity additive in the sense that taking two copies of the channel doubles its capacity? With regard to entanglement, additivity means that preparing two pairs of entangled particles gives twice the entanglement of one pair.

The output purity may be additive or multiplicative, depending on the purity measure of

# Chapter 4: Numerical Studies on the Additivity of Quantum Channel Capacities

choice. Since both terms describe the same behaviour, we will use the common expression 'additivity conjecture' while studying the multiplicativity properties of an output purity defined as follows:

$$\nu_p(\Phi) := \sup_\rho ||\Phi(\rho)||_p.$$

Here,

$$||\rho||_p = (\text{tr}\,|\rho|^p)^{1/p} \tag{4.3.1}$$

is the standard $p$-norm. $\nu_p$ is then multiplicative if

$$\nu_p(\Phi_1 \otimes \Phi_2) = \nu_p(\Phi_1)\nu_p(\Phi_2) \tag{4.3.2}$$

for arbitrary $\Phi_1$ and $\Phi_2$ and for $1 < p < \infty$. The tensor product of two channels acting on a bipartite state $\rho_{12}$ is defined as

$$(\Phi_1 \otimes \Phi_2)(\rho_{12}) := \sum_{i=1}^{L_1} \sum_{j=1}^{L_2} (K_i^{(1)} \otimes K_j^{(2)}) \rho_{12} (K_i^{(1)} \otimes K_j^{(2)})^\dagger. \tag{4.3.3}$$

Of particular interest is the limit $p \to 1$ which describes the additivity of classical channel capacities. A proof of additivity for this case would mean that we cannot transmit more classical information over multiple quantum channels if we use entangled states.

## 4.4 The Werner-Holevo channel as a counterexample

In [68], Werner and Holevo disproved the conjecture from Equation 4.3.2 for $p > 4.79$. As a counterexample, they used the following channel on $d \times d$ input matrices:

$$\begin{aligned}\Phi_{WH}(\rho) &:= \frac{1}{d-1}(\text{tr}(\rho)\mathbf{1}_d - \rho^t) \\ &= \frac{1}{2(d-1)} \sum_{i,j} (|i\rangle\langle j| - |j\rangle\langle i|)\rho(|i\rangle\langle j| - |j\rangle\langle i|)^\dagger\end{aligned} \tag{4.4.1}$$

Terms for the output purity of a single instance and for multiple copies of this channel can be derived analytically such that by simple numerics it could be shown in [68] that the additivity conjecture does not hold for $p > p_0 = 4.7823$.

CHAPTER 4: NUMERICAL STUDIES ON THE ADDITIVITY OF QUANTUM CHANNEL CAPACITIES

As one would expect, small variations of this channel, i.e.,

$$\tilde{\Phi}_{WH}(\rho) = (1-\epsilon)\Phi_{WH}(\rho) + \epsilon\,\Phi_X(\rho), \tag{4.4.2}$$

where $\Phi_X$ can be an arbitrary quantum channel and $\epsilon \in \mathbb{R}$ is small, are likely to also violate additivity. We studied this violation using for $\Phi_X$ the channels described in Section 4.2.3 and for a 'random unitary' channel whose Kraus operators are randomly chosen unitary matrices and whose Kraus rank was $d$.

As can easily be checked, the output purity of the Werner-Holevo channel is unitarily invariant, i.e., $\nu_p(\Phi_{WH}(\rho)) = \nu_p(\Phi_{WH}(U\rho U^\dagger))$, where $U$ is a unitary matrix. Furthermore, the channel has a flat spectrum of real eigenvalues $(1, \pm\frac{1}{2})$ and a determinant close to zero for $d > 2$. In the Bloch sphere picture, the channel leaves the $y$-components invariant while changing the signs of the $x$- and $z$-components.

## 4.5 Optimising the output purity using gradient flows

### 4.5.1 Statement of the problem

Can we find other channels that violate Equation 4.3.2? This question can be turned into an optimisation problem by setting $\rho = U\rho_0 U^\dagger$, where $\rho_0$ is a fixed initial state, and then find the maximum output purity for two copies of a candidate channel using a gradient flow on the unitary orbit of the initial state:

$$\nu_p(\Phi^{\otimes 2}) = \sup_{\rho_{12}} ||\Phi^{\otimes 2}(\rho_{12})||_p$$

$$= \sup_{\psi_{12} \in \mathbb{C}^{d^2}} ||\Phi^{\otimes 2}(|\psi_{12}\rangle\langle\psi_{12}|)||_p$$

$$= \sup_{U_{12} \in \mathbf{U}(d^2)} ||\Phi^{\otimes 2}(U_{12}|\psi^0_{12}\rangle\langle\psi^0_{12}|U^\dagger_{12})||_p.$$

Since the $p$-norm is a convex function which has its maximum at the extremal states, it suffices to consider pure states as inputs for this combined channel. Using Equation 4.3.1, we immediately see that essentially the term

$$F_p(\Phi \otimes \Phi, U_{12}) := \mathrm{tr}\big\{\big((\Phi \otimes \Phi)(U_{12}|\psi^0_{12}\rangle\langle\psi^0_{12}|U^\dagger_{12})\big)^p\big\} \tag{4.5.1}$$

CHAPTER 4: NUMERICAL STUDIES ON THE ADDITIVITY OF QUANTUM CHANNEL CAPACITIES

needs to be maximised. In order to violate the additivity conjecture, the term

$$\max F_p(\Phi \otimes \Phi, U_{12}) - \max F_p^2(\Phi, U_1)$$

must be positive, where $F(\Phi, U_1)$ represents the output purity for a single instance of the channel $\Phi$ (with $U_1 \in \mathbf{U}(d)$). Candidates for additivity-violating channels should be chosen from different classes of quantum channels (see below).

### 4.5.2 Description of the numerical procedure

We used the following iterative scheme to find other counterexamples to the additivity conjecture.

1. Set $|\psi^0\rangle = (1, 0, \ldots, 0) \in \mathbb{C}$.

2. Generate a random unitary matrix $U_1^{(0)} \in \mathbf{U}(d)$ from a distribution according to the Haar measure. Set $U_{12}^{(0)} = U_1^{(0)} \otimes U_1^{(0)}$.

3. Select as a channel either a known quantum channel or a random channel.

4. Set $k = 0$.

5. Using Equations 4.3.3 and 4.5.1, compute the quality function $F_p(\Phi, U_1^{(k)})$ for a single instance of the chosen channel, and $F_p(\Phi^{\otimes 2}, U_{12}^{(k)})$ for two copies. Stop if $F_p(\Phi^{\otimes 2}, U_{12}^{(k)}) - F_p^2(\Phi, U_1^{(k)})$ is positive.

6. Compute the gradient of $F$ with respect to $U$ according to

$$\nabla_U F_p := p \sum_{i=1}^{L} [\rho, K_i^\dagger \Phi^{p-1}(\rho) K_i]. \tag{4.5.2}$$

Here, we define $\rho$ as $\rho = U_1^{(k)} |\psi^0\rangle \langle \psi^0| (U_1^{(k)})^\dagger$ for the single channel and as $\rho = U_{12}^{(k)} |\psi^0 \psi^0\rangle \langle \psi^0 \psi^0| (U_{12}^{(k)})^\dagger$ for the combined channel. The $K_i$ are the respective Kraus operators. In order to derive this gradient, we used a Fréchet derivative, including the chain rule

$$\frac{\partial}{\partial X} H(G(X)) = \frac{d}{dY} H(Y) \cdot \frac{\partial}{\partial X} G(X),$$

where $Y = G(X)$, and the trace derivative

$$\frac{\partial}{\partial X} \text{tr}\{XAX^{-1}C\} = \{AX^{-1}C - X^{-1}CXAX^{-1}\}.$$

Note that this gradient is valid only for integer values of $p$ (see below).

7. Update $U_1$ and $U_{12}$ according to

$$U_1^{(k+1)} = \text{expm}\{-\gamma \nabla_U F_1(\Phi, U_1^{(k)})\} \cdot U_1^{(k)},$$
$$U_{12}^{(k+1)} = \text{expm}\{-\gamma \nabla_U F(\Phi^{\otimes 2}, U_{12}^{(k)})\} \cdot U_{12}^{(k)},$$

where $\gamma$ is a stepsize parameter.

8. Set $k \to k+1$ and go to step 5.

With a suitably chosen stepsize, this algorithm yields the maximal output purity for any channel given in its Kraus operator representation. We used a simple stepsize adjustment procedure that performed well in our setup: If the quality function $F$ increased during the last iteration, $\gamma$ is increased by 10% in the current iteration. If $F$ previously decreased, the last best choice for $U$ is taken and $\gamma$ is halved for the current iteration. The optimisation was stopped when the stepsize was smaller than the threshold value $10^{-8}$ or when the gradient norm was below $10^{-6}$.

One question naturally arises: What is the best dimension $d$ to look for new counterexamples? Based on the existing Werner-Holevo example and following [70], we studied the range $2 \leq d \leq 6$. For the exponent $p$, we chose a range of $2 \leq p \leq 30$. Only integer values could be used here as the matrix derivative yielding the gradient in Equation 4.5.2 is an open research problem for non-integer values. Depending on the channel, we found that a range of $2 \leq p \leq 1000$ was numerically feasible, but we focused on relatively small values of $p$ due to the significance of $p \to 1$ (see Section 4.3). The exact values of $d$ and $p$ for every channel are listed in Table 4.1.

As a starting point, we tested the Werner-Holevo channel and two hybrid versions according to Equation 4.4.2 where $\Phi_X$ was either a random unitary channel or a depolarising channel.

Next, we optimised the other channels listed in Table 4.1, including some standard channels and *extremal* random unitary channels, which we focused on. Extremal channels are extreme points in the convex set of quantum channels. A channel with Kraus operators $\{K_i\}$ is called extremal if the set of matrices $\{K_i^\dagger K_j\}$ is linearly independent [71]. Their Kraus operators can

CHAPTER 4: NUMERICAL STUDIES ON THE ADDITIVITY OF QUANTUM CHANNEL CAPACITIES

be constructed as

$$[K_i]_{\mu\nu} = \langle \mu i|(U|\nu 1)\rangle,$$

with $U$ being a (random) unitary matrix and $i, \mu, \nu = 1, 2, \ldots, d$. In our optimisations, we mainly used this type of channel to search for additivity violations since the standard channels are known to be additive [72, 73] and, at the time of this project, the Werner-Holevo channel was the only known counterexample.

Taking the extremal unitary channels, we pre-optimised them for a high initial depolarisation, thus mimicking this feature of the Werner-Holevo channel, before we started the gradient-based optimisation of the output purity. The pre-optimisation used the Matlab function fminsearch to find a set of $d$ unitary Kraus operators yielding a channel with a minimal initial value of $\nu_p(\Phi)$ with respect to a mutually unbiased basis. This channel then served as a starting point for the actual optimisation.

In total, we tested a few thousand channels (see Table 4.1 for details) using Matlab R2007a under Linux.

### 4.5.3 Results and further developments

| Channel | $d$ | $p$ | Optimisable? | Violation found? |
|---|---|---|---|---|
| bit flip | 2, 4, 6 | 2 - 20 | yes | no |
| phase flip | 2, 4, 6 | 2 - 20 | yes | no |
| depolarising | 2, 4, 6 | 2 - 20 | yes | no |
| amp. damping | 2, 4, 6 | 2 - 20 | yes | no |
| phase damping | 2, 4, 6 | 2 - 20 | yes | no |
| Casimir (see [74]) | 4 | 2 - 20 | yes | no |
| random unitary[1] | 3 - 6 | 5 - 30 | yes | no |
| Werner-Holevo | 3 - 6 | 5 - 30 | no[2] | yes |
| W-H/random hybrid | 3 - 6 | 5 - 10 | yes | yes[3] |
| W-H/depol hybrid | 4 | 5 - 20 | yes | yes[3] |

[1] incl. extremal channels    [2] unitarily invariant    [3] depends on $w$

**Table 4.1:** Additivity properties of the set of quantum channels that were used as test cases for the algorithm described in the previous section. Only the Werner-Holevo channel itself and small variations of it showed additivity violations for the dimensions and exponents that were studied.

Using the algorithm and the setup described in the previous section, we first verified the additivity violation of the Werner-Holevo channel and the hybrids derived from it. As men-

CHAPTER 4: NUMERICAL STUDIES ON THE ADDITIVITY OF QUANTUM CHANNEL
CAPACITIES

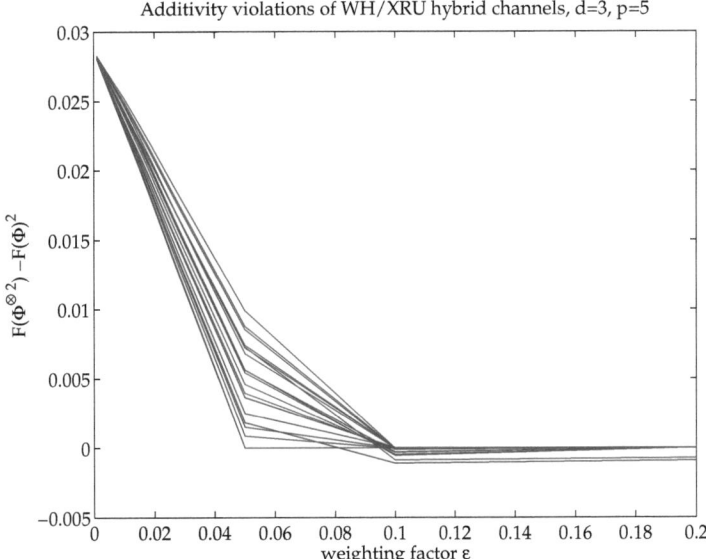

**Figure 4.5.1:** Additivity violations of 20 hybrid channels composed of Werner-Holevo and extremal random unitary (XRU) channels according to Equation 4.4.2. Optimisations were carried out using for $d = 3$ and $p = 5$. The strength of the violation depends on the weighting factor $\epsilon$. No violations are found for $\epsilon > 0.1$ in this case.

tioned before, the Werner-Holevo channel is invariant under unitary conjugation of its input and thus can not be optimised. For hybrid versions, however, we could numerically increase the output purity and detect additivity violations for small $\epsilon < \epsilon_0$. Figure 4.5.1 shows an example of violations for 20 hybrid channels (Werner-Holevo mixed with extremal random unitary channels) for $d = 3$ and $p = 5$. The exact values of $\epsilon_0$ vary with the dimension and with the exponent. These results imply that the Werner-Holevo channel does not represent a singularity in the 'search space' of quantum channels; we rather observe a smooth transition to channels that do not violate additivity. Table 4.1 shows the parameters and results for the Werner-Holevo channel and two examples of hybrids (mixing with the depolarising channel and with a random unitary channel). We note that if additivity is violated for $p = p_0$ in these cases, then the violation occurs for all $p > p_0$ if $d$ is constant. When a channel shows an additivity violation and we then increase the dimension $d$, $p$ must also be increased to preserve the violation.

Our search for new counterexamples in the class of (extremal) random unitary channels did

CHAPTER 4: NUMERICAL STUDIES ON THE ADDITIVITY OF QUANTUM CHANNEL CAPACITIES

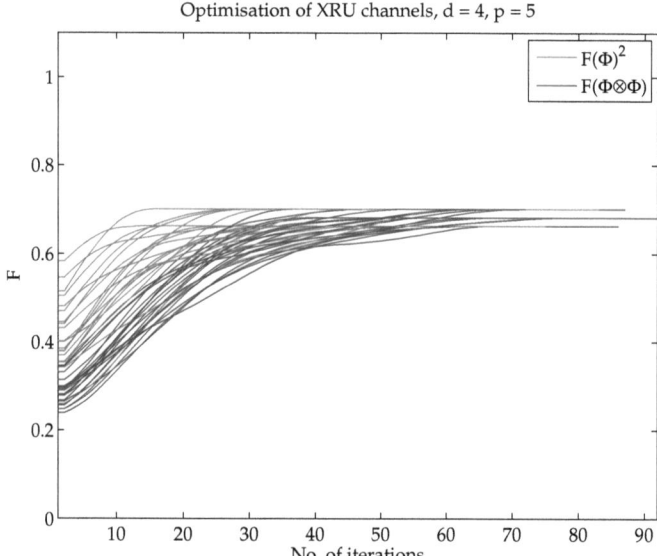

**Figure 4.5.2:** 20 optimisations of an extremal random unitary (XRU) channel with $d = 4$ and $p = 5$. The orange lines represent the value of $F(\Phi, U_1)$ of a single XRU channel, the blue lines represent the corresponding values of $F(\Phi^{\otimes 2}), U_{12}$ for two copies of the same channel. No additivity violations are found.

not succeed. This means, we were not able to find any additivity-violating channel in this class for $3 \leq d \leq 6$ and $5 \leq p \leq 30$. As an example, some optimisations of this class of channels are depicted in Figure 4.5.2 for $d = 4$ and $p = 5$. The 1000s of tested channels represent only a small portion of the search space of all possible channels, in particular because we restricted the optimisation to the given values of $d$ and $p$ to ensure numerical feasibility. Our search strategy of pre-optimising the channels to mimic properties of the Werner-Holevo channel (e.g., high initial depolarisation and small determinant) did not solve this fundamental problem.

In 2008, two publications [75, 76] disproved the additivity conjecture. Hayden and Winter [75] presented nonconstructive counterexamples for $1 \leq p \leq \infty$, explicitly excluding the case $p = 1$. In their proof, they rely on unitary channels in high dimensions ($d \to \infty$) for which they show the existence of additivity-violating channels without giving explicit examples. Unfortunately, the techniques developed in our work cannot be used to derive constructive counterexamples since the dimensions are beyond numerical tractability. The

CHAPTER 4: NUMERICAL STUDIES ON THE ADDITIVITY OF QUANTUM CHANNEL CAPACITIES

authors of [75, 76] stress that their findings do not have implications on the additivity of the minimal von Neumann entropy. Based on this work, Hastings [76] derived more general results for arbitrary $p$, using finite-dimensional random unitary channels of the form

$$\Phi(\rho) = \sum_{i=1}^{L} q_i U_i \rho U_i^\dagger,$$

where $q_i \geq 1$, $\sum q_i = 1$, $U \in \mathbf{U}(d)$, and $1 \ll L \ll d$. Again, we see that these counterexamples are channels in very high dimensions, unlike the ones we took as test cases for our optimisations. As Hayden and Winter, Hastings provided non-constructive examples, proving only the existence of such channels. His findings also hold for the additivity of the minimal von Neumann entropy and have thus a fundamental significance for quantum information theory.

CHAPTER 5

# Benchmarking a concurrent-update optimal-control algorithm

> Data is zeroes and ones. Software is zeroes and ones and hard work.
>
> Greg Wilson

## 5.1 Introduction

### 5.1.1 Overview

In this chapter, the idea of optimal control is introduced and applied to quantum systems. An optimal-control algorithm is presented that searches for pulse sequences to steer finite-dimensional quantum systems in an optimal way. It is known as the Gradient Ascent Pulse Engineering (GRAPE) algorithm [50]. This terminology, however, will not be used in the following as changes to the original proposal have introduced the usage of second-order information, whereas the term *gradient ascent* is sometimes used as a synonym for *steepest ascent* which is a first-order method. We use *concurrent-update algorithm* instead.

GRAPE has arisen from optimisation techniques in nuclear magnetic resonance (NMR) spectroscopy. Here, radio frequency pulses steer nuclear spin states of molecular or atomic ensembles. NMR spectroscopy is ideally suited for applying optimised pulses since it offers advanced technology for shaping RF pulses and provides relatively long coherence times. The principles behind GRAPE, however, are independent of the physical system to be optimised.

CHAPTER 5: BENCHMARKING A CONCURRENT-UPDATE OPTIMAL-CONTROL ALGORITHM

The algorithm maximises a given quality function subject to the equation of motion of the quantum system. To achieve this goal, the algorithm optimises the available control fields based on the first- and second-order derivative of the quality function. Typically, one chooses to optimise the vector of pulse amplitudes but an optimisation can also be done on the duration of a pulse or on the phase between pulses. Concurrent update means that all pulses in the sequence are changed in every iteration of the algorithm, unlike in sequential-update schemes [77, 78, 79].

The optimisation can be regarded as a gradient flow on the unitary group of Hamiltonian quantum dynamics. In the following, only closed quantum systems will be considered, but the algorithm works for open quantum systems with dissipative dynamics as well ([50, 80]).

### 5.1.2 Organisation

This chapter is organised as follows. Section 5.2 introduces the optimal control framework, which is applied to a quantum setting in Section 5.3. The concurrent-update algorithm is presented in detail in Section 5.4. The various options for computing gradients are discussed in Section 5.5. Another crucial module of the algorithm is the update method; Section 5.6 describes three of these methods. Another type of algorithm, a sequential-update scheme with a possible extension to a hybrid algorithm, is briefly presented in Sections 5.7 and 5.8. The chapter closes with numerical studies comparing the configurations discussed before.

## 5.2 The optimal control framework

The objective of optimal control is the optimisation of a dynamic system by finding controls that achieve a given optimality criterion. This dynamic system is typically described by a time-dependent state vector $x(t)$ that is manipulated by controls $u(t)$ over the time interval $[0, T]$. The optimality criterion is described by a scalar objective functional $\Phi$ of the form

$$\Phi = \Psi\big(x(T)\big) + \int_0^T L\big(x(t), u(t)\big) dt. \tag{5.2.1}$$

In this equation, $\Psi$ depends only on the state at the final time $T$, whereas the integral term represents a running cost. The control problem is to maximise $\Phi$ subject to the equation of

CHAPTER 5: BENCHMARKING A CONCURRENT-UPDATE OPTIMAL-CONTROL ALGORITHM

motion of the system, i.e.,

$$\max_u \Phi(x, u), \text{ subject to } \dot{x}(t) = f\Big(x(t), u(t)\Big) \qquad (5.2.2)$$

Here, $x(0) = x_0$ and $u(t)$ is restricted to the set of permissible controls. Note that, in many cases, this problem has multiple solutions. A time-optimal solution maximises $\Phi$ for a minimum value of $T$.

## 5.2.1 Pontryagin's maximum principle

Pontryagin's maximum principle provides a necessary condition for maximising the quality function 5.2.1. It can be derived by introducing a Lagrange multiplier vector $\lambda(t)$ into the problem. Assuming only real vectors, we define

$$h = \lambda^t f + L,$$

with $\lambda^t$ denoting the transpose of $\lambda$. If the variation of $h$ vanishes, we have found a necessary but not sufficient condition for global optimality. Therefore, in general only local extrema can be expected. A sufficient condition for the variation in $h$ to be zero is provided by the following criteria:

$$\frac{\partial h}{\partial u} = 0,$$
$$h(T) = 0,$$
$$\frac{d\lambda}{dt} = -\frac{\partial h}{\partial x},$$
$$\frac{dx}{dt} \equiv f = \frac{\partial h}{\partial \lambda^t}.$$

In general, these equations require numerical solutions as no analytic solutions can be found. Only in special cases (see e.g. [21]), analytic approaches exist.

For a more detailed introduction into optimal control, see the original work by Pontryagin et al. [18] and the book by Kirk [16].

## 5.3 Optimal control for quantum systems

Following Section 3.2, a closed quantum system is defined by its drift Hamiltonian $H_0$ and the control Hamiltonians $H_m$ corresponding to the real-valued control amplitudes $u_m$:

$$H_{tot}(t) := H_0 + \sum_{m=1}^{M} u_m(t) H_m \qquad (5.3.1)$$

The system can be driven externally by changing $u_m$. One obtains a bilinear control system whose dynamics are governed by the Schrödinger equation:

$$|\dot{\psi}(t)\rangle = -i H_{tot}(t) |\psi(t)\rangle . \qquad (5.3.2)$$

If one chooses to neglect the unobservable global phase or treat dissipative systems, the density operator representation of Equation (5.3.2) can be taken:

$$\dot{\rho}(t) = -i[H_{tot}(t), \rho(t)]. \qquad (5.3.3)$$

This equation is also known as the Liouville-von Neumann equation. With $U(t) := \exp\{-it H_{tot}(t)\}$ and $|\psi(t)\rangle = U(t)|\psi(0)\rangle$, Equation (5.3.2) can be lifted to the operator level and thus becomes independent of the initial and final states $|\psi(0)\rangle$ and $|\psi(T)\rangle$:

$$\dot{U}(t) = -i H_{tot}(t) U(t) \qquad (5.3.4)$$

The control problem for synthesising a target operator, or quantum gate, $U_G$ can now be described as

$$\max_{u} \Phi(u), \text{ subject to } \dot{U}(t) = -i H_{tot}(t) U(t) \qquad (5.3.5)$$

Remember from Equation 5.2.1 that, in the general case, the quality function $\Phi$ contains a term for the quality at the final time $T$ and a term representing the running cost. An example of the latter is the power the system consumes during the propagation from the inital to the target state. The final quality reflects the distance between the achieved final state $|\psi(T)\rangle$ and the target state $|\psi\rangle_G$, or the achieved unitary operation $U(T)$ and the desired gate $U_G$. In the following, the quality term will only consist of the final quality and we will focus on

# CHAPTER 5: BENCHMARKING A CONCURRENT-UPDATE OPTIMAL-CONTROL ALGORITHM

state-independent optimisations of a target operator $U_G$.

Note that in this setting, any optimisation algorithm will generally find only local extrema. Using a large set of initial conditions is one way to increase the chances of finding global extrema instead of local ones. Another, more sophisticated method for achieving this goal is *tabu search* [81, 82]. It combines a technique for leaving local extrema with a list of already visited points in the search space. In tests performed with tabu search, however, we did not find better results than by simply choosing many initial conditions, which is our method of choice in the following studies.

## 5.3.1 The quality function

There are two ways for expressing the geometrical distance between the two unitary operators $U_G$ and $U(T)$. The first one takes the global phase between the two operators into account and yields the quality term

$$\Phi_1 := \operatorname{Re} \operatorname{tr}\{U_G^\dagger U(T)\}/N, \tag{5.3.6}$$

which is normalised to 1. $\Phi_1$ achieves the maximum of 1 if and only if $U(T) = U_G$, which follows from

$$\|U(T) - U_G\|_2^2 = 2N - 2\operatorname{Re} \operatorname{tr}\{U_G^\dagger U(T)\} \tag{5.3.7}$$

and the property of the Hilbert-Schmidt norm $\|x\| = 0 \Leftrightarrow x = 0$. In this case, we optimise over the special unitary group $\mathbf{SU}(N)$, where $N$ is the matrix dimension.

For practical applications the global phase can be neglected. For this purpose we define a quality function that is insensitive to any global phase factors:

$$\Phi_2 := |\operatorname{tr}\{U_G^\dagger U(T)\}|^2/N^2. \tag{5.3.8}$$

This function is maximised if

$$U(T) = e^{-i\theta} U_G \tag{5.3.9}$$

for any $\theta \in [0, 2\pi]$. This yields an optimisation over the projective special unitary group $\mathbf{PSU}(N)$:

$$\mathbf{PSU}(N) \stackrel{\text{iso}}{=} \frac{\mathbf{SU}(N)}{\mathbb{Z}_N} \stackrel{\text{iso}}{=} \frac{\mathbf{U}(N)}{\mathbf{U}(1)}. \tag{5.3.10}$$

CHAPTER 5: BENCHMARKING A CONCURRENT-UPDATE OPTIMAL-CONTROL ALGORITHM

Expressions for quality functions in the case of state-to-state optimisations for pure and mixed states can be found in [50].

## 5.4 Algorithmic scheme

Here, we present an iterative scheme for performing the maximisation of Equation 5.3.5 by concurrently updating the control vector **u**. The following steps describe this algorithm. A pseudocode representation of this scheme is given in Algorithm 5.1. Note that we write $\Phi$ instead of $\Phi_1$ or $\Phi_2$ when the exact form of the quality function is irrelevant.

**0. Initial Setup** Fix a final time $T$ and a digitisation $K$ such that $T$ is divided into $K$ time steps $t_k$ with $\Delta t_k = t_k - t_{k-1} = T/K$. During each step $k$, the control term $H_m$ and the corresponding control amplitude $u_m(t_k)$ are constant. Choose a random initial value $u_m^{(0)}(t_k)$ for all $k$ and $m$.

**1. Compute Hamiltonians** Compute the total Hamiltonian

$$H_{tot}(t_k) = H_0 + \sum_m u_m(t_k) H_m \quad \forall\, t_k. \tag{5.4.1}$$

**2. Exponentiate** Obtain the propagators $U_k$ by computing the exponentials

$$U_k = e^{-i\Delta t_k H_{tot}(t_k)} \quad \forall\, k. \tag{5.4.2}$$

**3. Propagate Forward** Set $U_0 = \mathbf{1}$ and calculate the forward propagation

$$U_{k:0} := U_k \cdot U_{k-1} \cdots \cdots U_1 \cdot U_0 \quad \forall\, k. \tag{5.4.3}$$

**4. Propagate Backward** Similarly, set $U_{K+1} = U_G^\dagger$ and calculate the backward propagation

$$\lambda_{K+1:k+1} := U_{K+1} \cdot U_K \cdots \cdots U_{k+2} \cdot U_{k+1} \quad \forall\, k. \tag{5.4.4}$$

**5. Evaluate Quality** Evaluate the quality function according to Equation 5.3.6 or Equation 5.3.8,

$$\Phi^{(r)} = \operatorname{Re}\operatorname{tr}\{U_G^\dagger U(T)\}/N = \operatorname{Re}\operatorname{tr}\{\lambda_{K+1:k+1} U_{k:0}\}/N$$
$$\text{or } \Phi^{(r)} = |\operatorname{tr}\{U_G^\dagger U(T)\}|^2/N^2 = |\operatorname{tr}\{\lambda_{K+1:k+1} U_{k:0}\}|^2/N^2,$$

where $r$ is the index of the current iteration.

6. **Get Gradient** Compute the gradient vector $\nabla \Phi = \left( \frac{\partial \Phi(U_{1:0})}{\partial u_1(t_1)}, \ldots, \frac{\partial \Phi(U_{k:0})}{\partial u_m(t_k)}, \ldots, \frac{\partial \Phi(U_{K:0})}{\partial u_M(t_K)} \right)$ for all $t_k$ and $u_m$ using one of the formulae derived in Section 5.5. We introduce the shorthand notation $\nabla_{k,m}\Phi = \frac{\partial \Phi}{\partial u_m(t_k)}$.

7. **Update Controls** Update the control vector by one of the methods described in section 5.6.

8. **Check Stopping Criteria** Iterate steps 1 through 7 until $\Phi^{(r)} > 1 - \epsilon$. The goal tolerance $\epsilon$ is introduced because we cannot expect a numerical method to reach a final fidelity of exactly 1. In practice, several other stopping criteria are applied, e.g., an upper limit for the number of iterations, a lower limit for the norm of the gradient or the step, and a lower limit for the change of $\Phi$ from iteration $r$ to $r+1$.

---

**Algorithm 5.1** Pseudocode for the concurrent-update algorithm

Divide final time T into K time steps $t_k$.
Choose a (random) pulse sequence.
REPEAT
    FOR each $t_k$
        Compute total Hamiltonian.
        Compute propagator.
        Compute forward propagation.
        Compute backward propagation.
        Compute gradient.
    ENDFOR
    Evaluate quality function $\Phi$.
    Update pulse sequence.
UNTIL $\Phi \geq 1 - \epsilon$

---

Figure 5.4.1 shows a schematic representation of the concurrent-update algorithm applied to the $m$-th control vector. The gradient information in iteration $r$ is used to compute the amplitudes in all time slots $t_k$ for the next iteration $r+1$.

## 5.5 Gradient computation

### 5.5.1 The gradient formula with respect to the control amplitudes

The derivative of the quality function $\Phi_1$ with respect to $u_m(t_k)$ is

## CHAPTER 5: BENCHMARKING A CONCURRENT-UPDATE OPTIMAL-CONTROL ALGORITHM

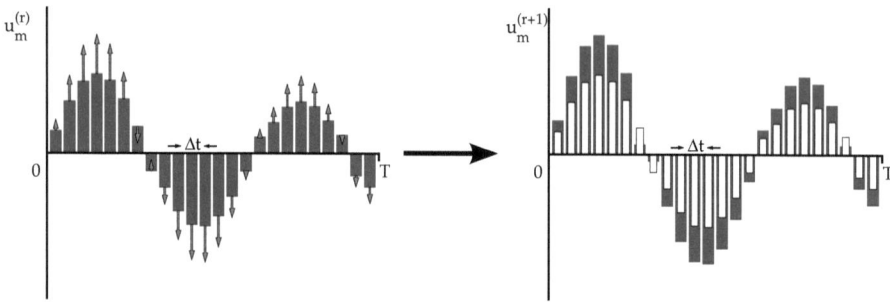

**Figure 5.4.1:** A simplified example of a concurrent-update algorithm. The vector of piecewise-constant control amplitudes $u_m$ (blue bars) in iteration $r$ is updated using the gradient information (red arrows) to give the new pulse sequence in iteration $r + 1$.

$$\begin{aligned}
\nabla_{k,m}\Phi_1 &= \operatorname{Re}\nabla_{k,m}\operatorname{tr}\{\lambda_{K+1:k+1}U_{k:0}\}/N \\
&= \operatorname{Re}\operatorname{tr}\{\nabla_{k,m}(\lambda_{K+1:k+1}U_{k:0})\}/N \\
&= \operatorname{Re}\operatorname{tr}\{\lambda_{K+1:k+1}(\nabla_{k,m}U_{k:0})\}/N \\
&= \operatorname{Re}\operatorname{tr}\{\lambda_{K+1:k+1}(\nabla_{k,m}U_k)U_{k-1:0}\}/N \\
&= \operatorname{Re}\operatorname{tr}\{\lambda_{K+1:k+1}(\nabla_{k,m}e^{-i\Delta t_k H_{tot}(t_k)})U_{k-1:0}\}/N.
\end{aligned} \quad (5.5.1)$$

Since the control Hamiltonians do not commute with the total Hamiltonian in the general case, calculating the derivative of the exponential is nontrivial. Four methods will be presented to obtain this derivative: an approximation of the gradient term to first order in $\Delta t_k$ (the standard approximation), the well-known finite-difference method, an approximation using a series expansion, and an exact method based on the eigendecompostion of the total Hamiltonian. All of these methods have different numerical demands.

#### 5.5.1.1 The standard gradient approximation

The general approach for computing the derivative of the exponential of a matrix function $f(x)$ is

$$\frac{\partial}{\partial x}e^{f(x)} = \int_0^1 e^{sf(x)}\frac{\partial f}{\partial x}e^{(1-s)f(x)}ds. \quad (5.5.2)$$

CHAPTER 5: BENCHMARKING A CONCURRENT-UPDATE OPTIMAL-CONTROL ALGORITHM

This follows from Equation I.8 in Reference [83] together with the usual definition of a derivative. From this we arrive at

$$\frac{\partial U_k}{\partial u_m(t_k)} = -i\left(\int_0^{\Delta t_k} U_k(\tau) H_m U_k(-\tau) d\tau\right) U_k, \tag{5.5.3}$$

with

$$U_k(\tau) = \exp\{-i\tau H_{tot}(t_k)\}.$$

For a small enough $\Delta t_k$ the unitary terms in the integral of Equation (5.5.3) can be expanded to first order in $\tau$. We obtain

$$\int_0^{\Delta t_k} U_k(\tau) H_m U_k(-\tau) \approx \int_0^{\Delta t_k} \left(\mathbf{1} - i\tau H_{tot}(t_k)\right) H_m \left(\mathbf{1} + i\tau H_{tot}(t_k)\right) d\tau$$

$$\approx \int_0^{\Delta t_k} H_m - i\tau[H_{tot}(t_k), H_m] d\tau. \tag{5.5.4}$$

This approximation requires

$$\Delta t_k \ll ||H_{tot}(t_k)||_2^{-1} \quad \forall k \tag{5.5.5}$$

in order to be valid. By computing the integral and dropping the $\Delta t^2$ term we get

$$\int_0^{\Delta t_k} U_k(\tau) H_m U_k(\tau) \approx \Delta t H_m. \tag{5.5.6}$$

Thus, a first order approximation of the derivative of $U_k$ with respect to $u_m(t_k)$ is

$$\frac{\partial U_k}{\partial u_m(t_k)} \approx -i\Delta t_k H_m U_k, \tag{5.5.7}$$

which yields the gradient expression for the quality function $\Phi_1$:

$$\nabla_{k,m} \Phi_1 = -\mathrm{Re}\,\mathrm{tr}\{\lambda_{K+1:k+1} i\Delta t_k H_m U_{k:0}\}/N. \tag{5.5.8}$$

Similarly, we have the following approximate gradient term for the quality function $\Phi_2$:

$$\nabla_{k,m} \Phi_2 = -2\mathrm{Re}\,\mathrm{tr}\{\lambda_{K+1:k+1} i\Delta t_k H_m U_{k:0}\}\mathrm{tr}\{U_{k:0}\lambda_{K+1:k+1}\}/N^2. \tag{5.5.9}$$

### 5.5.1.2 Computing the exponential derivative by an eigendecomposition

We can use the following lemma by Aizu [84] to *exactly* compute the derivative of the exponential term in Equation (5.5.1):

**Lemma 4.** *The derivative of the exponential of a sum of two non-commuting operators A and xB with respect to x at x = 0 is given by*

$$\langle \xi_\mu | \frac{d}{dx} e^{A+xB} | \xi_\nu \rangle \bigg|_{x=0} = \begin{cases} \langle \xi_\mu | B | \xi_\nu \rangle e^{\xi_\mu} & \text{if } \xi_\mu = \xi_\nu \\ \langle \xi_\mu | B | \xi_\nu \rangle \frac{e^{\xi_\mu} - e^{\xi_\nu}}{\xi_\mu - \xi_\nu} & \text{else} \end{cases} \quad (5.5.10)$$

*where the vectors $|\xi_\mu\rangle$ denote the eigenvectors of the operator A and the coeffcicients $\xi_\mu$ denote the eigenvalues of that operator: $A|\xi_\mu\rangle = \xi_\mu |\xi_\mu\rangle$.*

A proof of this lemma is given in Appendix B.

Inserting this into Equation (5.5.1) yields the gradient

$$\nabla_{k,m} \Phi_1 = \text{Re tr} \{ \tilde{\lambda}_{K+1:k+1} D_{k,m} \tilde{U}_{k-1:0} \} / N, \quad (5.5.11)$$

where the elements of $D_{k,m}$ are computed according to Equation (5.5.10) (with $A = -i\Delta t H_{tot}(t_k)$ and $B = -i\Delta t_k H_m$) and the operators $\lambda_{K+1:k+1}$ and $U_{k-1:0}$ are transformed into the eigenbasis $\xi_\mu$.

Similarly, for the quality function $\Phi_2$ we have the following exact gradient expression:

$$\nabla_{k,m} \Phi_2 = 2\text{Re tr}\{\lambda'_{K+1:k+1} D_{k,m} U'_{k-1:0}\} \text{tr}\{(U'_{k:0})^\dagger (\lambda'_{K+1:k+1})^\dagger\} / N^2. \quad (5.5.12)$$

This method of computing the gradient comes with an extra advantage: The evaluation of the matrix exponential becomes trivial since only the diagonal matrix of eigenvalues needs to be exponentiated.

Note that the method as described here cannot be applied when optimising an open system. The Liouvillian operator $\mathcal{L} = iH + \Gamma$ does not satisfy $\mathcal{L}^\dagger \mathcal{L} = \mathcal{L}\mathcal{L}^\dagger$ in the generic case, so a derivation of the right-hand side of Equation 5.5.10 becomes nontrivial as the eigenvectors to different eigenvalues are not necessarily orthogonal to each other (see Appendix B).

### 5.5.1.3 Computing the exponential derivative by a finite-difference method

The derivative of a general function $f$ at a point $x$ is defined by the limit

$$\frac{\partial}{\partial x} f = \lim_{\epsilon \to 0} \frac{f(x+\epsilon) - f(x)}{\epsilon}. \tag{5.5.13}$$

When $\epsilon$ is a fixed non-zero value, the fraction on the right-hand side is an approximation of the derivative of $f$. In our case, Equation (5.5.13) becomes

$$\nabla_{k,m} \Phi_1 = \frac{\operatorname{Re} \operatorname{tr}\{\lambda_{K+1:l+1} P_{k,m} U_{k-1:0}\}/N - \Phi_1}{\epsilon}, \tag{5.5.14}$$

with

$$P_{k,m} = \exp\{-i\Delta t_k (H_{tot}(t_k) + \epsilon H_m)\}. \tag{5.5.15}$$

For $\Phi_2$ we find

$$\nabla_{k,m} \Phi_2 = \frac{|\operatorname{tr}\{\lambda_{K+1:k+1} P_{k,m} U_{k-1:0}\}|^2 / N^2 - \Phi_2}{\epsilon}. \tag{5.5.16}$$

Numerically, this derivative can be regarded as exact to machine precision when $\epsilon$ is sufficiently small. In practice, however, $\epsilon$ cannot be made arbitrarily small without facing numerical instabilities. Typically, one finds the best results when choosing $\epsilon$ to be on the order of $10^{-7}$. Figure 5.5.1 compares finite-difference gradients with exact gradients computed according to the previous subsection.

### 5.5.1.4 Computing the exponential derivative by a Hausdorff series

The derivative of the exponential can also be expressed as a Hausdorff series,

$$\frac{\partial}{\partial x} \exp\{A + xB\}|_{x=0} = \exp\{A\} \left( B + \frac{[B,A]}{2} + \frac{[[B,A],A]}{6} + \ldots \right), \tag{5.5.17}$$

that can be computed to machine precision. While this procedure can be cumbersome in the general case, it is particularly efficient when computing with large sparse matrices, when a small number of terms is sufficient to yield a high accuracy. See Reference [85] for details.

CHAPTER 5: BENCHMARKING A CONCURRENT-UPDATE OPTIMAL-CONTROL ALGORITHM

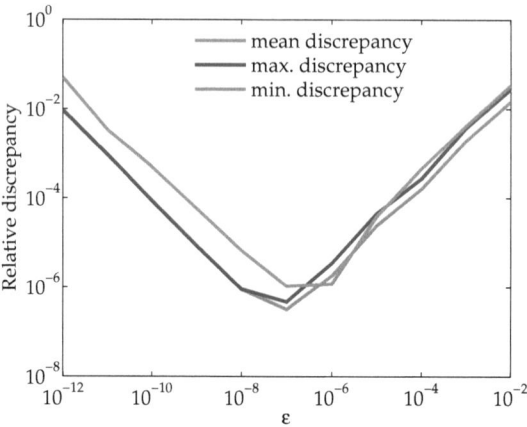

**Figure 5.5.1:** Accuracy of the finite-difference gradient method as a function of the parameter $\epsilon$. The relative discrepancy is defined as $\mathrm{abs}(\nabla_{ex} - \nabla_{fd})/\mathrm{abs}(\nabla_{ex})$, where $\nabla_{ex}$ is the exact gradient and $\nabla_{fd}$ is the finite-difference gradient. The gradients were calculated under the 32 bit Linux version of Matlab R2010a.

### 5.5.2 The gradient formula with respect to time

Apart from optimising over the piecewise constant control amplitudes, one can choose the durations of the $k$ timeslices $\Delta t_k = t_k - t_{k-1}$ as control parameters [86]. In this case, the derivative of $\exp\{-i\Delta t_k H_{tot}(t_k)\}$ can be calculated trivially since the derivative of the exponent commutes with the derivative of the exponential function, i.e.,

$$[-iH_{tot}(t_k), \exp\{-i\Delta t_k H_{tot}(t_k)\}] = 0. \quad (5.5.18)$$

With the shorthand notation $\nabla_k = \frac{\partial}{\partial \Delta t_k}$, we then obtain

$$\nabla_k \Phi_1 = -\mathrm{Re}\,\mathrm{tr}\{\lambda_{K+1:k+1}(iH_{tot}(t_k)U_{k:0})\}/N \quad (5.5.19)$$

and

$$\nabla_k \Phi_2 = -\mathrm{tr}\{\lambda_{K+1:k+1}(iH_{tot}(t_k)U_{k:0})\}\mathrm{tr}\{U_{k:0}\lambda^\dagger_{K+1:k+1}\}/N^2 + c.c.$$
$$= -2\mathrm{Re}\,\mathrm{tr}\{\lambda_{K+1:k+1}(iH_{tot}(t_k)U_{k:0})\}\mathrm{tr}\{U_{k:0}\lambda^\dagger_{K+1:k+1}\}/N^2. \quad (5.5.20)$$

CHAPTER 5: BENCHMARKING A CONCURRENT-UPDATE OPTIMAL-CONTROL ALGORITHM

## 5.5.3 The gradient formula with respect to phase

In a typical NMR control setting, one can find pulses for controlling the x- and y-magnetisation of each spin [87, 88]. The amplitude $u_x$ is related to $u_y$ by a phase $\theta$ and an overall pulse amplitude $u_0$, as shown in Figure 5.5.2:

$$u_x(\theta) = u_0 \cos(\theta) \quad (5.5.21)$$
$$u_y(\theta) = u_0 \sin(\theta) \quad (5.5.22)$$

When $u_0$ is fixed, the controls $u_x$ and $u_y$ are functions of the phase alone. For an

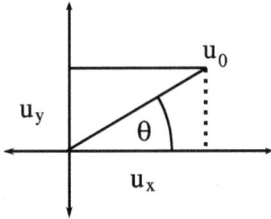

**Figure 5.5.2:** Phase relation between an x- and a y-pulse in a typical NMR setting. The overall pulse amplitude is $u_0$.

optimisation over all $M \cdot K$ phases $\theta_m(t_k)$ the gradient can be derived by starting from the general Equation (5.5.1) and setting

$$H_{tot}(t_k) = H_0 + \sum_{m=1}^{M} u_x\big(\theta_m(t_k)\big) H_{x,m} + u_y\big(\theta_m(t_k)\big) H_{y,m} \quad (5.5.23)$$

$$= H_0 + \sum_{m=1}^{M} u_{m,0}(t_k) \cos\big(\theta_m(t_k)\big) H_{x,m} + u_{m,0}(t_k) \sin\big(\theta_m(t_k)\big) H_{y,m} \quad (5.5.24)$$

Along the lines of Equation (5.5.11) it follows that

$$\nabla_{k,m} \Phi_1 = \operatorname{Re} \operatorname{tr}\{\tilde{\lambda}_{K+1:k+1} D'_{k,m} \tilde{U}_{k-1:0}\}/N,$$

where the elements of $D'_{k,m}$ are computed according to Equation (5.5.10) with

$$A = -i\Delta t_k H_{tot}(t_k)$$

and

$$B = -i\Delta t_k\, u_{m,0}(t_k) \Big\{ -\sin\big(\theta_m(t_k)\big) H_{x,m} + \cos\big(\theta_m(t_k)\big) H_{y,m} \Big\}.$$

For $\Phi_2$, we find the gradient

$$\nabla_{k,m}\Phi_2 = 2\operatorname{Re}\operatorname{tr}\{\lambda'_{K+1:k+1}\tilde{D}_{k,m}U'_{k-1:0}\}\operatorname{tr}\{(U'_{k:0})^\dagger(\lambda'_{K+1:k+1})^\dagger\}/N^2. \quad (5.5.25)$$

## 5.6 Update methods

There exist many methods for updating the control vector in an iterative optimisation scheme. The following three methods have been shown to be of particular importance for optimal control algorithms. In principle, they can be divided into first- and second-order approaches, depending on their use of gradient and possibly Hessian information.

### 5.6.1 Steepest ascent

Steepest ascent is a first-order method for computing the control vector for the next iteration. Only the gradient information is used to determine the new vector. In the case of our pulse optimisation, the control amplitudes in the next iteration are

$$u_m^{(r+1)}(t_k) = u_m^{(r)}(t_k) + \gamma^{(r)}\nabla_{k,m}\Phi. \quad (5.6.1)$$

Here, $\gamma^{(r)}$ is a stepsize parameter that can be found using a line search. Thus, it typically changes from one iteration to the next. For small enough $\gamma^{(r)}$, $\Phi^{(r+1)} > \Phi^{(r)}$ is guaranteed. Equation (5.6.1) illustrates the resemblance to Euler's method for solving ordinary differential equations.

Steepest ascent shows a slow convergence in many practical cases, especially when the function has elongated valleys, i.e., when the problem is poorly scaled. One then observes zig-zag or hemstitching patterns near the optimum [89].

### 5.6.2 Newton and quasi-Newton methods

This method uses second-order information by taking the Hessian matrix into account when computing the new set of control parameters for the next iteration. The update rule is

$$u^{(r+1)}(t_k) = u^{(r)}(t_k) + \gamma^{(r)}(\operatorname{Hess}^{-1}\nabla\Phi)_k, \quad (5.6.2)$$

CHAPTER 5: BENCHMARKING A CONCURRENT-UPDATE OPTIMAL-CONTROL ALGORITHM

where $\gamma^{(r)}$ is again a stepsize parameter. Here, to simplify notation and without loss of generality, we consider the case $M = 1$, i.e., only one control Hamiltonian. Using the Hessian information improves the convergence behaviour near the optimum but increases the computational demands as computing this matrix is costly. This drawback of Newton methods can be overcome by employing an approximation procedure for the Hessian, in which case one speaks of a *quasi-Newton method*. The most popular of these methods is BFGS, which will be discussed in the following.

#### 5.6.2.1 Approximating the Hessian with BFGS and l-BFGS

The Broyden-Fletcher-Goldfarb-Shanno (BFGS) method is a quasi-Newton approach that merely needs first-order information to approximate the Hessian matrix $B$. $B$ is updated at every iteration $r$ according to

$$B_{r+1} = B_r + \frac{y_r y_r^t}{y_r^t \Delta x_r} - \frac{B_r \Delta x_r \Delta x_r^t B_r}{\Delta x_r^t B_r \Delta x_r},$$

where $y_r = \nabla f(x_{r+1}) - \nabla f(x_r)$

and $\Delta x_r = x_{r+1} - x_r$.

The limited-memory variant of BFGS (l-BFGS) is particularly well suited for optimisation problems with a large number of dimensions, as the Hessian never actually needs to be computed or stored. Instead, a history of the last $j$ updates of $x$ and $\nabla f(x)$ is remembered. Typically, the method performs well even when this history is short, e.g., on the order of 10 iterations. This knowledge about previous runs is used to do operations implicitly that would require the Hessian or its inverse. In many cases, $H_0 = \mathbf{1}$ is chosen as a starting point such that the first step is equivalent to a gradient ascent.

For an extensive analysis of the BFGS and the l-BFGS method, see [25].

### 5.6.3 Conjugate gradients

Combining the strengths of the first- and second-order methods described above is the main idea behind the conjugate-gradient method [90]. Far away from the optimum, this algorithm behaves like a first-order method. When approaching the optimum, its behaviour changes to that of a second-order method, although the Hessian is never computed.

Two steps characterise this algorithm:

1. At the current point, a sequence of conjugate (or orthogonal) directions $d_1, \ldots d_{K \cdot M}$ is created.

2. The optimum in each direction is determined. It is the starting point for the search in the next direction:
$$u_m^{(r+1)}(t_k) = u_m^{(r)}(t_k) - \alpha_i d_i, \qquad (5.6.3)$$
with $\alpha_i = \arg\min \Phi(u_m^{(r)}(t_k) - \alpha_i d_i)$ being a scalar that is to be determined by a one-dimensional search. Supposing that $\{d_i\}$ forms an orthogonal basis, we obtain
$$u_m^{(N+1)}(t_k) = u_m^{(1)}(t_k) - \sum_{i=1}^{M \cdot K} \alpha_i d_i. \qquad (5.6.4)$$

It can be shown [15] that $u_m^{(N+1)}(t_k)$ will be the optimum if $\Phi$ is quadratic, and it will be a good approximation to the optimum if $\Phi$ is not quadratic.

### 5.6.4 Line search

The three methods presented above share the need of finding the optimal value of a parameter ($\gamma$ or $\alpha$) along a given direction. This procedure is called a *line search*. Many approaches for implementing this line search exist, but a full study of these is beyond the scope of this work. In the numerical studies described in Section 5.9, we relied on toolbox functions provided by the software packages we used. Their default settings were assumed to combine line search strategies and update methods in an optimal way.

## 5.7 Comparison with a sequential-update algorithm

A sequential-update algorithm following the works of Krotov [79, 77, 78] can be used as an alternative to a concurrent-update algorithm. In practice, this algorithm comes in many flavours that are usually referred to as "Krotov methods". The common and characteristic property of all these methods is that only one time slot is updated per iteration. In the following, we will describe a method developed by Schirmer et al. [24].

Algorithm 5.2 shows the pseudocode for the sequential-update algorithm by Schirmer.

CHAPTER 5: BENCHMARKING A CONCURRENT-UPDATE OPTIMAL-CONTROL ALGORITHM

| **Algorithm 5.2** Pseudocode for the sequential-update algorithm |
|---|
| Divide final time T into K time steps $t_k$. |
| Choose a (random) pulse sequence. |
| REPEAT |
|     FOR each $t_k$ |
|         Compute forward propagation $U_{k-1:0} := U_{k-1} \ldots U_0$. |
|         Compute backward propagation from $\lambda_{K+1:k+1} := U_{K+1} \ldots U_{k+1}$. |
|         Compute gradient. |
|         Update controls at $t_k$. |
|         Compute updated total Hamiltonian $H_{tot}(t_k)$. |
|         Compute updated propagator $U_k$. |
|     ENDFOR |
|     Evaluate quality function $\phi$. |
| UNTIL $\phi \geq 1 - \epsilon$ |

One immediately verifies that, per iteration, only two matrix multiplications are needed for the forward and backward propagation. Since $U_{k:0} = U_k \cdot U_{k-1:0}$ and $\lambda_{K+1:k} = \lambda_{K+1:k+1} \cdot U_k$, the existing product of matrices from the last iteration can be used for computing the propagation terms in the current iteration. In this implementation, the control vector at $t_k$ is updated only once before the algorithm proceeds to the next time slot. Other implementations, however, perform several updates, depending on the gradient norm or the change of the quality function.

In comparison with the concurrent-update case, the small numerical cost per iteration comes at the price of a higher number of iterations needed to reach the goal quality. Furthermore, the gradient needed for the exponential in time slot $k$ requires an update in time slot $k - 1$. Thus, an extra matrix exponential occurs in the sequential-update algorithm, in addition to the eigendecomposition needed for the gradient (when using an exact gradient). The concurrent-update algorithm merely needs the eigendecomposition which provides the matrix exponential at no extra cost (see Section 5.5.1.2).

The gradient can be computed according to the procedures described in Section 5.5. Finding the best way to update the control vector, however, is currently an open problem. Among the three principal methods from Section 5.6, a BFGS-based method is the least favourable. As each iteration updates another set of controls, the Hessian in the current iteration cannot be approximated using previous Hessian information. While BFGS is well-matched with the concurrent-update algorithm, a fast second-order method for the sequential algorithm

CHAPTER 5: BENCHMARKING A CONCURRENT-UPDATE OPTIMAL-CONTROL ALGORITHM

is yet to be found.

For our numerical studies (see below), we thus used a gradient-ascent update scheme that was the fastest method available for this implementation of a Krotov-type algorithm[1]. Its performance depends crucially on the step-size parameter $\gamma^{(r)}$. Although choosing a small constant $\gamma^{(r)}$ ensures convergence (to a critical point of the quality function) this is usually a very bad choice. Conventional step-size control algorithms also appear to be inefficient, as they require revaluation of the objective function for each trial step size. Even in the sequential-update scenario, this requires the evaluation of a matrix exponential and a Hadamard product. For this reason, we based our step size control on a simple heuristic. Assuming we can locally approximate the quality function $\Phi$ by a quadratic function in the step-size parameter $\gamma^{(r)}$ along the gradient direction,

$$\Phi(\gamma^{(r)}) = \gamma^{(r)}(2 - \gamma^{(r)}),$$

the linear approximation is $2\gamma^{(r)}$ and the error term is $(\gamma^{(r)})^2$. Our step-size control is based on trying to ensure that the actual gain in the fidelity is about 50% of the expected gain (being $2\gamma^{(r)}$)}, which is attained at the maximum of this simple model. Thus, we start with an initial guess for the step size $\gamma^{(r)}$ and evaluate $\Phi(\gamma^{(r)})$. If the increase $\Delta\Phi = \Phi(\gamma^{(r)}) - \Phi(0)$ is less than $\frac{2}{3}$ of the expected gain, then the step size was too large and we decrease $\gamma^{(r)}$ by a small factor (0.99 for the following runs). If the actual gain $\Delta\Phi$ is greater than $\frac{4}{3}$ of the expected gain, then the step size was too small and we increase $\gamma^{(r)}$ by a small factor (e.g., 1.01). Rather than applying this change for the current time step, which would require reevaluating the fidelity, we apply it only in the next time step, i.e.,

$$\gamma^{(r+1)} = \begin{cases} 0.99 \cdot \gamma^{(r)} & \text{if } \Delta\Phi < \frac{4}{3}\gamma^{(r)} \\ 1.01 \cdot \gamma^{(r)} & \text{if } \Delta\Phi > \frac{8}{3}\gamma^{(r)} \end{cases}. \quad (5.7.1)$$

For the sequential-update algorithm with many time steps, avoiding the computational overhead of multiple fidelity evaluations is usually preferable compared to the small gain that could be achieved by adjusting the step size $\gamma^{(r)}$ at the current time step. Starting with $\gamma^{(r=0)} = 1$ as a default value, we found that $\gamma^{(r)}$ usually quickly converges to an optimal (problem-specific) value and only varies very little after this initial adjustment period.

---

[1]This description is a replication of a paragraph written by Sophie Schirmer from the University of Cambridge for an upcoming collaborative publication.

## 5.8 Hybrid algorithms

The concurrent- and the sequential-update algorithm represent the two extremes of a set of possible algorithms. In between those ends, hybrid schemes allow for optimising subsets of timeslots. Many kinds of hybrids can be defined by how these subsets are chosen and possibly changed between iterations. In order to reduce the number of expensive matrix operations, a smart management of the propagation terms is required, which creates more overhead than in the simple cases of concurrent- and sequential-update algorithms. The major drawback, though, is the lack of a suitable method for the update-step. The range of possible hybrid algorithms is likely to require a range of update methods, each adapted to the respective algorithm. As this is an open research problem, hybrid algorithms will not be discussed in the remainder of this work.

## 5.9 Numerical studies

### 5.9.1 Test environment

For the numerical studies presented in this chapter, the following test configuration was used, if not stated otherwise.

All optimisations were carried out under the Linux version of Matlab R2009b in 64 bit single-thread mode. The CPU was one core of an AMD Opteron Dual Core @2.6 GHz that could access 8 GB of main memory. This machine represented one node of a high-performance computing cluster at the *Leibniz-Rechenzentrum München*.

The concurrent-update algorithm was implemented using the optimisation toolbox in Matlab. For this purpose, the initial maximisation problem from Section 5.3 was transformed into a minimisation problem by changing the signs of the quality and gradient functions. As a consequence, steepest ascent will be steepest descent in the following. The fminunc function allowed for the realisation of a BFGS update module with a cubic line search procedure for unconstrained problems. We will only consider problems without constraints unless we explicitly give constraints, e.g. upper and lower bounds on the control vector. For these examples, the toolbox function fmincon was used. It is based on the interior-point algorithm described in Appendix D.

When comparing running times of optimisation algorithms, we were interested in a program's total running time from start to completion. For this purpose, we chose to record the wall time instead of the CPU time. The former includes the time spent on communication and on saving and loading data. The Matlab commands *tic* and *toc* were used to measure wall time.

We set the goal quality to $10^{-4}$ and the tolerances for the change in the control, function, or gradient vectors to $10^{-8}$. For all test problems, the operation times were sufficiently long to ensure the problems were solvable with full quality. We did not aim for time-optimal solutions, for which we expect a similar behaviour.

For statistically significant results, we repeated every measurement for 20 random initial control vectors that had a mean value of 0 and a standard deviation of 1.

### 5.9.2 Toy models used for numerical optimisations

The following three toy models were used as test problems for all numerical optimisations. The goal was to cover a variety of systems in terms of dimension, coupling topology, coupling type, and spin quantum number. All systems were fully controllable in the sense of Chapter 3.

Two types of optimisations were performed with these systems: unitary gate synthesis and state-to-state transfers. Both can be treated in the same formalism following equations 5.4.3 and 5.4.4. For state-to-state transfers, one simply substitutes $U_0$ and $U_G$ by column vectors. As only closed quantum systems were studied in this work, all states were pure states.[2] The initial states were random pure states; the final states were the product of the input state and the target gate for the respective model.

#### 5.9.2.1 Model 1: a spin chain

The first model is a chain of three Heisenberg-coupled spins which are individually controllable by x- and y-controls. The set of Hamiltonians is thus:

$$H_0 = \sum_{\mu=x,y,z} J_{12}^\mu \sigma_\mu^{(1)} \sigma_\mu^{(2)} + J_{23}^\mu \sigma_\mu^{(2)} \sigma_\mu^{(3)}$$

---

[2]The formalism works for mixed states in essentially the same way, by using the superoperator formalism (see Appendix A) to turn density matrices into vectors. This squares the dimension of all matrices but allows for computing matrix-vector products instead of the numerically more demanding matrix-matrix products.

CHAPTER 5: BENCHMARKING A CONCURRENT-UPDATE OPTIMAL-CONTROL ALGORITHM

$$H_{1,2} = \frac{1}{2}\sigma_{x,y}^{(1)}$$
$$H_{3,4} = \frac{1}{2}\sigma_{x,y}^{(2)}$$
$$H_{5,6} = \frac{1}{2}\sigma_{x,y}^{(3)}.$$

Here, $\sigma_\mu^i$ denotes the Pauli matrix $\sigma_\mu$ (\mu=x,y,z) at spin $i$, and $J_{ij}^\mu$ denotes the $\mu$-th coupling term between spins $i$ and $j$. In agreement with the nomenclature in Chapter 3, this type of spin chain is referred to as an ABC chain. All coupling constants are set to 1. The target gate is a 3-spin quantum Fourier transformation (QFT). For state-to-state transfer optimisations, the initial state is a random state vector and the target state is the target gate applied to this initial state.

### 5.9.2.2 Model 2: a maximally coupled spin network for cluster state preparation

As an example of a fully-coupled spin-$\frac{1}{2}$ model, we studied a system of four locally addressable spins interacting via the Ising-ZZ coupling:

$$H_0 = \frac{J}{2}\sum_{i=1}^{3}\sum_{j=i+1}^{4} \sigma_z^i \sigma_z^j$$
$$H_{1,2} = \frac{1}{2}\sigma_{x,y}^1$$
$$H_{3,4} = \frac{1}{2}\sigma_{x,y}^2$$
$$H_{5,6} = \frac{1}{2}\sigma_{x,y}^3$$
$$H_{7,8} = \frac{1}{2}\sigma_{x,y}^4$$

This system can be used to prepare a cluster state by applying the quantum gate $U_{CS} = \exp(-i\pi/2 H_{cs})$ to the state $|\psi_0\rangle = ((|0\rangle + |1\rangle)/\sqrt{2})^{\otimes 4}$. Here, the effective Hamiltonian

$$H_{CS} = \frac{J}{2}(\sigma_z^1\sigma_z^2 + \sigma_z^2\sigma_z^3 + \sigma_z^3\sigma_z^4 + \sigma_z^4\sigma_z^1)$$

represents a $C_4$ graph. Again, the coupling constant $J$ is set to 1. $U_{CS}$ is used as the target gate. For optimisations of state-to-state transfers, $|\psi_0\rangle$ is used as the initial state and the cluster state as the target.

CHAPTER 5: BENCHMARKING A CONCURRENT-UPDATE OPTIMAL-CONTROL ALGORITHM

### 5.9.2.3 Model 3: a driven 7-level system

Optimisations were also carried out on a test system with the following drift and control Hamiltonians:

$$H_0 = J_z^2$$
$$H_1 = J_z$$
$$H_2 = J_x.$$

The $J_i$ are the total angular momentum operators corresponding to the quantum number $j = 3$. We choose a random unitary $7 \times 7$ matrix as the target gate for unitary optimisations, and random initial and target states for state-to-state optimisations.

## 5.9.3 Comparison of gradient methods

### 5.9.3.1 Parameter set

In this subsection, the four gradient methods presented in Section 5.5.1 are compared with respect to their relative speed and to the final quality they were able to achieve in optimisations.

We present the setup and results for model 1. We chose the final time $T = 5$ sec when optimising for a unitary gate, and $T = 0.4$ sec when optimising a state-to-state transfer. These were not the smallest times for this problem, but they were of the same order of magnitude and sufficiently small to make the optimisations nontrivial. In order to test the dependence of the standard approximation on the relation between $\Delta t_k$ and $||H_{tot}(t_k)||_2$ according to Equation (5.5.5), we chose different values for the number of time slots and amplitudes of the initial pulse sequence. These parameters and the corresponding values of the product $P := \text{mean}\left(\Delta t_k \cdot ||H_{tot}(t_k)||_2\right)$ are listed in Table 5.1, where we denote the standard deviation of the initial pulse sequence by $s$, $s := \text{std}(u_{ini}(t_k))$.

CHAPTER 5: BENCHMARKING A CONCURRENT-UPDATE OPTIMAL-CONTROL ALGORITHM

Unitary optimisation: values of $P = \text{mean}\left(\Delta t_k \cdot ||H_{tot}(t_k)||_2\right)$. ($T = 5$ s)

| s \ K | 50 | 100 | 150 | 200 | 500 |
|---|---|---|---|---|---|
| 0.1 | 0.2067 | 0.1034 | 0.0689 | 0.0517 | 0.0207 |
| 10 | 1.9201 | 0.9661 | 0.6465 | 0.4840 | 0.1932 |

State-to-state optimisation: values of $P = \text{mean}\left(\Delta t \cdot ||H_{tot}(t_k)||_2\right)$. ($T = 0.4$ s)

| s \ K | 20 | 40 | 50 | 70 | 80 | 100 |
|---|---|---|---|---|---|---|
| 10 | 0.3846 | 0.1929 | 0.1533 | 0.1105 | 0.0960 | 0.0769 |
| 100 | 3.7279 | 1.8882 | 1.5044 | 1.0768 | 0.9411 | 0.7509 |

**Table 5.1:** Values of $P$ for the different numbers of time slots $K$ and initial amplitudes which are represented by the standard deviation of the initial pulse sequence (denoted by s). The top table shows the values for the unitary optimisations, the bottom one shows the values for the optimisations of state-to-state transfers.

We then performed three types of measurements:

1. In optimisations of model 1, we measured the quality as a function of the wall time for all four gradient methods, see Figure 5.9.1.

2. We measured the final quality reached using the standard approximation as a function of $P$ (Figure 5.9.2).

3. For the standard approximation, the difference between the approximated and the exact gradient vector was studied as a function of $P$. Since the gradients of only the first iteration needed to be compared, no actual optimisations were performed. We therefore chose a broader range of values of $P$ than given in Table 5.1 (see Figure 5.9.3), and studied models 2 ($T = 14$ sec) and 3 ($T = 1$ sec) in addition to model 1.

The first two measurements answer questions with a direct practical importance, whereas the last measurement is the most straight-forward way to assess the (theoretical) quality of the standard approximation.

For the finite-difference method, we set $\epsilon = 10^{-7}$. The Hausdorff series was cut off when the norm of the next term was below $10^{-13}$. The current quality $q$ and the elapsed time were

CHAPTER 5: BENCHMARKING A CONCURRENT-UPDATE OPTIMAL-CONTROL ALGORITHM

recorded at each iteration of an optimisation.

### 5.9.3.2  Results

Figure 5.9.1 shows the deviation from the maximum quality, $\Phi = 1$, as a function of the wall time for each gradient method and for a sample set of values of the product $P$ in the case of unitary optimisations (state-to-state transfer optimisations yield similar results). The thick lines represent the average of all runs with one method.

One observes that the standard approximation (pink lines) is competitive to the exact gradient method (blue lines) in terms of speed if it reaches or comes close to the goal quality, i.e., if the approximation is valid. This is the case when the digitisation is high, see Figure 5.9.1 (c) and (d). In cases (a) and (b), the standard approximation breaks down and cannot achieve qualities much higher than 0.99. In the same cases, the timing differences between the exact gradients on one hand and the finite-difference method and the Hausdorff series on the other hand are higher than in cases (c) and (d), where all four methods perform similarly fast. It is noted that the finite-difference method and the Hausdorff series reach the goal quality in all cases.

In Figure 5.9.2 we depict the mean final deviation from $\Phi = 1$ achieved by the exact gradient method (dashed blue lines) and by the standard approximation (orange and green lines), depending on the value of $P$. A unitary and a state-to-state optimisation of model 1 are shown. Note that in these examples the optimisation changed the standard deviation of the pulse sequence only to a small extent, i.e., $P_{initial} \approx P_{final}$. It can be seen that the approximation breaks down with increasing $P$ in all cases. The breakdown points vary among unitary and state-to-state optimisations, but $P$ needs to be at least one order of magnitude smaller than 1 to ensure the goal quality is reached when using the standard approximation. The exact method yields qualities that match the goal in all runs.

The same measurements as decribed above were performed for models 2 and 3. They yield similar outcomes, with one exception: the differences in running times between the gradient methods can be more pronounced for large K, with the exact gradients being ahead of all other methods; see Appendix C for results for model 2.

The accuracy of the standard approximation can also be assessed in a more direct way: by measuring the difference between the approximated and the exact gradient vector as a function of $P$. Thus, with $grad := \nabla_{k,m}\Phi$, we compute the vector $|grad_{approx} - grad_{exact}|/|grad_{exact}|$ (for 20 repetitions) and take its mean value. Figure 5.9.3 shows the measured values as a

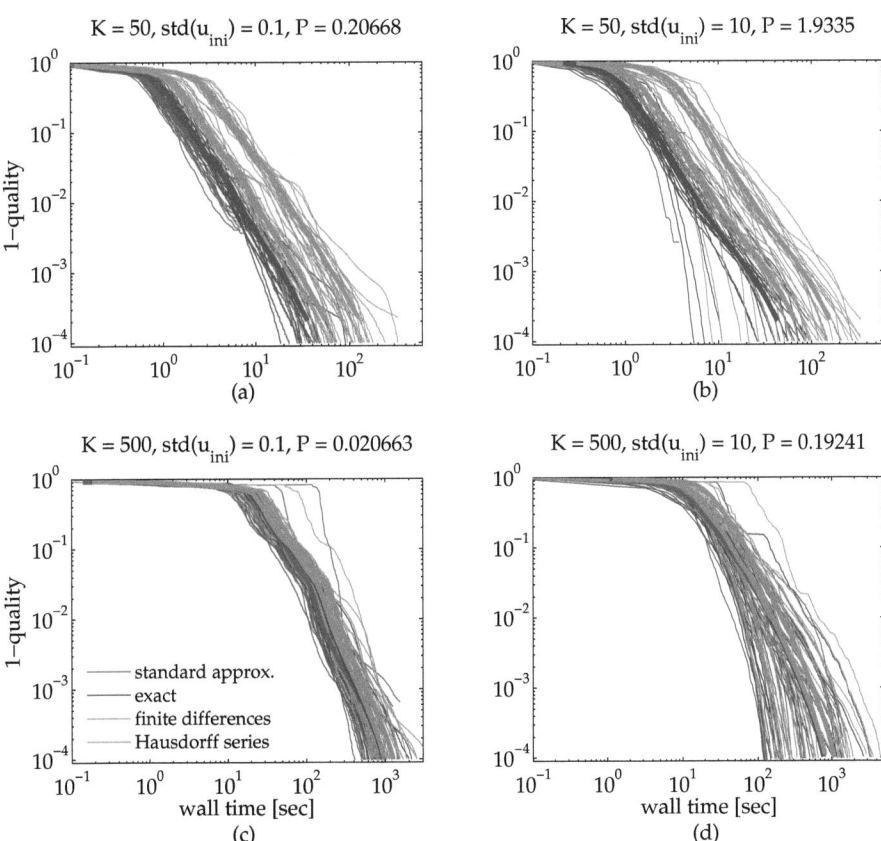

**Figure 5.9.1:** Doubly logarithmic plot showing the performance of the four gradient methods discussed in Section 5.5 for the optimisation of a unitary gate in model 1. The thick lines represent average values. The exact gradient method (blue line) performes best in all cases. For the small digitisation in cases (a) and (b), we observe a significant breakdown of the standard approximation (pink). Furthermore, the finite-difference gradient method (green) and the Hausdorff series (orange) are slower than the exact procedure. When a high digitisation is used in cases (c) and (d), the standard approximation achieves higher goal qualities, and all approximative methods perform as fast as the exact method.

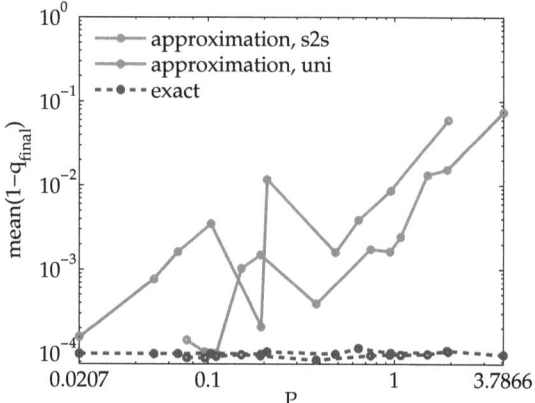

**Figure 5.9.2:** Doubly logarithmic plot showing the mean final deviations from $\Phi = 1$ as a function of $P$ for model 1. The exact gradient method (blue) converged to the goal quality in almost all cases, whereas the standard approximation (orange and green) did so only when $P$ was small. Note that the breakdown of the standard approximation depends on the type of optimisations (*s2s* denotes a state-to-state transfer, *uni* denotes a unitary optimisation), but $P$ is best chosen to be at least one order of magnitude smaller then 1.

function of $P$, where for $P$, again, the mean over all time slots $t_k$ was taken. As mentioned before, we took models 1-3 into account for this comparison, and studied a broader range of values for $P$.

It becomes clear that one needs to go to $P < 10^{-2}$ in order to make sure the error in the standard approximation is below 10%. As Figure 5.9.3 illustrates, the error depends on the model and can differ by an order of magnitude for the same $P$. For all models, however, a decrease of $P$ by one order of magnitude results in an error reduction by also one order of magnitude. If $P \approx 1$, the error of the standard approximation is typically around 100%.

Note that the accuracies of the finite-difference method and the Hausdorff method are independent of $P$. They depend only on the stepsize $\epsilon$ (see Figure 5.5.1) and the cut-off term, respectively.

### 5.9.3.3 Discussion

The exact gradient procedure involves an eigendecomposition with a numerical complexity of $\mathcal{O}(N^3)$. The two main advantages of this diagonalisation are (i) a perfectly accurate gradient and (ii) the lack of any extra matrix exponentials as we need to exponentiate scalars

CHAPTER 5: BENCHMARKING A CONCURRENT-UPDATE OPTIMAL-CONTROL ALGORITHM

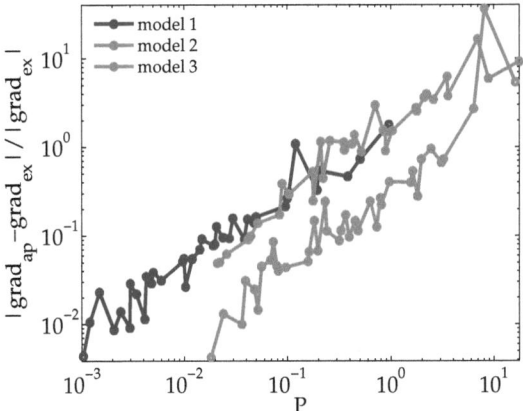

**Figure 5.9.3:** Relative difference between the exact gradient and the approximate gradient for different values of $P$ in optimisations of models 1 (blue), 2 (orange), and 3 (green). The logarithmic y-axis shows the mean values of the difference $|\text{grad}_{approx} - \text{grad}_{exact}|/|\text{grad}_{exact}|$, the logarithmic x-axis shows the mean values of $P$.

only, see Equation (5.5.10).

The finite-difference method also has an extra numerical cost of $\mathcal{O}(N^3)$, in this case for the matrix exponential, but it provides a gradient that is less accurate, depending on the choice of the stepsize $\epsilon$. This explains why it performs slightly worse than the exact method, although reaching the goal quality in all tests, see Figure 5.9.1 . Its advantage is the ability to provide a gradient when no other approximation or exact method is known.

When the standard approximation fails, as is the case in Figure 5.9.1(a) and (b), either the digitisation is chosen too small or the norm of the total Hamiltonian is too large. The gradient vector then points into a wrong direction of the search space such that eventually no minimum can be found. From Figure 5.9.2, there seems to be no fixed value at which the approximation breaks down. It rather depends on the the type of optimisation and the model to be optimised. As a rule of thumb, a value of $P < 0.1$ is required for the approximation to hold. When the approximation is valid, this gradient method performs as fast as the exact method. The gradient quality certainly has a strong influence on the outcome of an optimisation, but it does not solely determine the success of an optimisation, as a comparison of Figures 5.9.2 and 5.9.3 shows: even a value of $P$ that yields a gradient error of the order of 10% can give satisfactory results in an optimisation.

The Hausdorff series can be computed without any eigendecompositions or matrix expo-

CHAPTER 5: BENCHMARKING A CONCURRENT-UPDATE OPTIMAL-CONTROL ALGORITHM

nentials, as it relies merely on matrix multiplications. These have the same complexity $\mathcal{O}(N^3)$, but in numerically small systems like our example, matrix multiplications are faster in practice. Still, the number of series terms needed for a decent accuracy is too high to be competitive with the exact or finite-difference gradient methods. In this regard, the Hausdorff series seems to be tailored to larger spin systems with sparse matrix operations where one can expect a performance boost [91, 92, 85].

### 5.9.4 Comparison of update methods

#### 5.9.4.1 Parameters

Here, we present optimisations on models 1 and 3. The operation time for model 1 was set to $T = 10$ sec, the digitisation to $K = 100$. For model 3, we chose $T = 15$ sec and $K = 150$. We allowed a maximum of 5,000 iterations. The exact method from Section 5.5.1.2 was used for computing the gradients.

As the Matlab toolbox function `fminunc` does not allow the direct application of a simple conjugate-gradient method, we instead used the Matlab package `minFunc` by Mark Schmidt [93]. The options *sd*, *cg*, and *lbfgs* were chosen for the respective update methods. The steepest-descent method used a line search strategy that was based on an Armijo backtracking with cubic interpolation from new function and gradient values. The other two methods were combined with a bracketing strategy with cubic interpolation and extrapolation with function and gradient values. These were the software's default settings and thus assumed to be optimal choices.

#### 5.9.4.2 Results

Figure 5.9.4 shows the deviation from the maximum quality, $\Phi = 1$, as a function of the wall time for each update method. Figure 5.9.4 (a) illustrates the results for model 1, Figure 5.9.4 the results for model 3. The thick lines represent the average of all runs with one method. The quasi-Newton method and the conjugate-gradient method succeeded to reach the goal quality for both cases, whereas the steepest-descent method failed in case (b). When it converged in case (a), it was an order of magnitude slower than the conjugate gradient method. The fastest convergence was achieved by the quasi-Newton method.

CHAPTER 5: BENCHMARKING A CONCURRENT-UPDATE OPTIMAL-CONTROL ALGORITHM

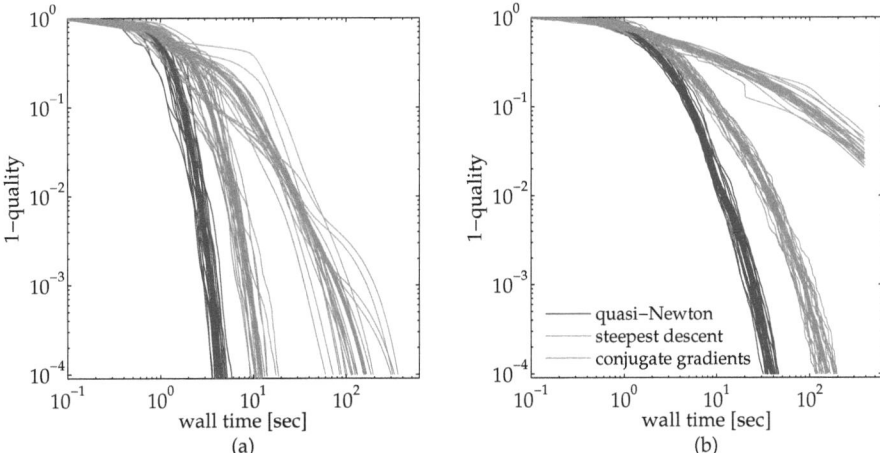

**Figure 5.9.4:** Comparison of three update methods for model 1 (a) and for model 3 (b). The quasi-Newton update method (blue lines) outperformed the conjugate-gradient method (orange) by reaching the goal quality in less than half of the wall time, on average. The steepest-descent method (green) was either an order of magnitude slower, see case (a), or it clearly failed to yield high qualities, see case (b).

#### 5.9.4.3 Discussion

In summary, using second-order information with the l-BFGS approach outperformed the purely gradient-based methods. Despite its additional cost of computing an approximation of the Hessian, l-BFGS converged substantially faster than the other two methods. The first-order steepest descent method was the least favourable of the three methods due to its bad convergence behaviour. Using conjugate gradients could speed up the optimisations significantly, yet was still inferior to the quasi-Newton approach.

#### 5.9.4.4 Other systems and optimisations of state-to-state transfers

In optimisations for model 2, we obtained the same qualitative results as for model 3. Another set of tests was carried out in which a state-to-state transfer with pure states was optimised. The results were similar to those for optimisations for unitary gates, with l-BFGS being the strongest method, but the performance differences to the other methods were less pronounced.

CHAPTER 5: BENCHMARKING A CONCURRENT-UPDATE OPTIMAL-CONTROL ALGORITHM

## 5.9.5 Comparison of sequential- and concurrent-update algorithms

### 5.9.5.1 Specification of test problems

Comparing two algorithms requires a more detailed analysis than comparing two modules of an algorithm, as we did in the previous sections. Therefore, we extended our set of test problems: we studied the 23 systems listed in Table 5.2 as test problems for our optimisation algorithms. This test suite included spin chains, a system of completely coupled trapped ions, an NV-centre system, and two driven spin-$j$ systems with $j = 3, 6$. We tried to cover many systems of practical importance with a range of coupling topologies and control schemes. We used large sets of parameters like system size, final time, number of time slots, and target gates. While it was impossible to include all types of control problems, we believe our subset of problems is comprehensive for the purpose of this comparison of algorithms.

**Spin chains:** Problems 1-12 are Ising-ZZ spin chains of various length $n = 1,\ldots,5$, in which the spins were addressable by individual $x$- and $y$-controls. The Hamiltonians for these systems take the following form:

$$H_0 = \tfrac{J}{2} \sum_{i=1}^{n-1} \sigma_z^i \sigma_z^{i+1}$$
$$H_{1,2} = \tfrac{1}{2} \sigma_{x,y}^1$$
$$H_{3,4} = \tfrac{1}{2} \sigma_{x,y}^2$$
$$\vdots$$

where $J = 1$. In the notation of Chapter 3, these are $ABC\ldots$ spin chains.

Only in problem 1, we consider linear crosstalk (e.g., via off-resonant excitation), leading to the control Hamiltonians

$$H_{1,2} = \alpha_{1,2}\sigma_x^1 + \alpha_{2,1}\sigma_x^2$$
$$H_{3,4} = \beta_{2,1}\sigma_y^1 + \beta_{1,2}\sigma_y^2$$

where $\alpha_c$ and $\beta_c$ are crosstalk coefficients. We chose $\alpha_1 = \beta_2 = 1$ and $\alpha_2 = \beta_1 = 0.1$.

Test problems 17 and 18 represent five-qubit Ising chains. A local Stark shift term is added to the drift Hamiltonian $H_0$. The control consists of simultaneous $x$- and $y$-rotations on all

**Table 5.2:** Test problems for comparing a sequential- and a concurrent-update scheme.

| Problem | Quantum System | Matrix Dimensions | K | T [1/J] | Target Gate |
|---|---|---|---|---|---|
| 1 | $AB$ Ising-ZZ chain | 4 | 30 | 2 | CNOT |
| 2 | $AB$ Ising-ZZ chain | 4 | 40 | 2 | CNOT |
| 3 | $AB$ Ising-ZZ chain | 4 | 128 | 3 | CNOT |
| 4 | $AB$ Ising-ZZ chain | 4 | 64 | 4 | CNOT |
| 5 | $ABC$ Ising-ZZ chain | 8 | 120 | 6 | QFT |
| 6 | $ABC$ Ising-ZZ chain | 8 | 140 | 7 | QFT |
| 7 | $ABCD$ Ising-ZZ chain | 16 | 128 | 10 | QFT |
| 8 | $ABCD$ Ising-ZZ chain | 16 | 128 | 12 | QFT |
| 9 | $ABCD$ Ising-ZZ chain | 16 | 64 | 20 | QFT |
| 10 | $ABCDE$ Ising-ZZ chain | 32 | 300 | 15 | QFT |
| 11 | $ABCDE$ Ising-ZZ chain | 32 | 300 | 20 | QFT |
| 12 | $ABCDE$ Ising-ZZ chain | 32 | 64 | 25 | QFT |
| 13 | $C_4$ Graph-ZZ | 16 | 128 | 7 | $U_{CS}$ |
| 14 | $C_4$ Graph-ZZ | 16 | 128 | 12 | $U_{CS}$ |
| 15 | NV-center | 4 | 40 | 2 | CNOT |
| 16 | NV-center | 4 | 64 | 5 | CNOT |
| 17 | $AAAAA$ Ising-ZZ chain | 32 | 1000 | 125 | QFT |
| 18 | $AAAAA$ Ising-ZZ chain | 32 | 1000 | 150 | QFT |
| 19 | $ABCDE$ Heisenberg chain | 32 | 300 | 30 | QFT |
| 20 | $A00$ Heisenberg-XXX chain | 8 | 64 | 15 | rand U |
| 21 | $AB00$ Heisenberg-XXX chain | 16 | 128 | 40 | rand U |
| 22 | driven spin-6 system | 13 | 100 | 15 | rand U |
| 23 | driven spin-3 system | 7 | 50 | 5 | rand U |

spins:

$$H_0 = \frac{J}{2}\sum_{i=1}^{5} \sigma_z^i \sigma_z^{i+1} - (i+2)\sigma_z^i$$

$$H_1 = \frac{1}{2}\sum_{i=1}^{5} \sigma_x^i$$

$$H_2 = \frac{1}{2}\sum_{i=1}^{5} \sigma_{x'}^i$$

Problem 19 is a Heisenberg-XXX coupled chain of 5 spins with a global always-on fields inducing simultaneous $x$-rotations on all spins:

$$H_0 = \frac{J}{2}\sum_{i=1}^{4}(\sigma_x^i \sigma_x^{i+1} + \sigma_y^i \sigma_y^{i+1} + \sigma_z^i \sigma_z^{i+1} - 10\sigma_x^i).$$

Control is provided by local Stark shift terms

$$H_i = \sigma_z^i \quad (i = 1,\ldots,5).$$

In problems 20 and 21, the spins are coupled by the Heisenberg-XXX interaction and the chains are subject to $x$- and $y$-controls on only one end (one or two spins, respectively):

$$H_0 = \frac{J}{2}\sum_{i=1}^{n-1} \sigma_x^i \sigma_x^{i+1} + \sigma_y^i \sigma_y^{i+1} + \sigma_z^i \sigma_z^{i+1}$$

$$H_{1,2} = \tfrac{1}{2}\,\sigma_{x,y}^1$$

$$(H_{3,4} = \frac{1}{2}\,\sigma_{x,y}^1)$$

Here, $J = 1$ and $n = 3, 4$. Restricting the controls in this way makes the systems harder to control and thus raises the bar for the optimisation.

**Completely coupled spin network for cluster state preparation:** The effective Hamiltonian of test problems 13 and 14,

$$H_{CS} = \tfrac{J}{2}(\sigma_z^1 \sigma_z^2 + \sigma_z^2 \sigma_z^3 + \sigma_z^3 \sigma_z^4 + \sigma_z^4 \sigma_z^4),$$

represents a $C_4$ graph of Ising-ZZ coupled qubits which can be used for cluster state preparation according to [94]. The underlying physical system is a completely Ising-coupled set of 4 ions that each represented a locally addressable qubit:

$$H_0 = \frac{J}{2} \sum_{i=1}^{3} \sum_{j=i+1}^{4} \sigma_z^i \sigma_z^j$$

$$H_{1,2} = \frac{1}{2} \sigma_{x,y}^1$$

$$H_{3,4} = \frac{1}{2} \sigma_{x,y}^2$$

$$H_{5,6} = \frac{1}{2} \sigma_{x,y}^3$$

$$H_{7,8} = \frac{1}{2} \sigma_{x,y}^4$$

Again, the coupling constant $J$ was set to 1. The following unitary is chosen as a target gate, which applied to the state $|\psi_1\rangle = ((|0\rangle + |1\rangle)/\sqrt{2})^{\otimes 4}$ generats a cluster state:

$$U_G = \exp(-i\frac{\pi}{2} H_{CS}).$$

**NV-centres in diamond:** In test problems 15 and 16, we optimise for a CNOT gate on two strongly coupled nuclear spins at an nitrogen-vacancy (NV) center in diamond, as described in [95]. In the eigenbasis of the coupled system, after a transformation into the rotating frame, the Hamiltonians are of the form:

$$H_0 = \text{diag}(E_1, E_2, E_3, E_4) + \omega_c \text{ diag}(1, 0, 0, -1)$$
$$H_1 = \tfrac{1}{2}(\mu_{12}\sigma_{12}^x + \mu_{13}\sigma_{13}^x + \mu_{24}\sigma_{24}^x + \mu_{34}\sigma_{3,4}^x)$$
$$H_2 = \tfrac{1}{2}(\mu_{12}\sigma_{12}^y + \mu_{13}\sigma_{13}^y + \mu_{24}\sigma_{24}^y + \mu_{34}\sigma_{34}^y)$$

Here, $E_1 \ldots E_4$ are the energy levels, $\omega_c$ is the carrier frequency of the driving field and $\mu_{\alpha,\beta}$ is the relative dipole moment of the transition between levels $\alpha$ and $\beta$. We choose the following values for our optimisations: $\{E_1, E_2, E_3, E_4\} = 2\pi\{-134.825, -4.725, 4.275, 135.275\}$ MHz, $\omega_c = 2\pi \times 135$ MHz, $\{\mu_{12}, \mu_{13}, \mu_{24}, \mu_{34}\} = \{1, 1/3.5, 1/1.4, 1/1.8\}$.

# CHAPTER 5: BENCHMARKING A CONCURRENT-UPDATE OPTIMAL-CONTROL ALGORITHM

**General driven multi-level systems:** As a candidate for a non spin-1/2 system, in test problems 22 and 23 we consider a Hamiltonian of the form

$$H_0 = J_z^2$$
$$H_1 = J_z$$
$$H_2 = J_x,$$

where the $J_i$ are total angular momentum operators. The $J_z^2$ term represents the drift Hamiltonian, the other two terms function as controls. We chose $j = 6$ for problem 22 and $j = 3$ for problem 23.

### 5.9.5.2 Numerical Details

As Table 5.2 shows, we optimised each test system for one of four quantum gates: a controlled-NOT (CNOT), a quantum Fourier transformation, a random unitary, or a unitary for cluster state preparation. The random unitary gate was found to be numerically more demanding than the other gates. The final times $T$ were always chosen sufficiently long to ensure the respective problem was solvable with full fidelity. Hence, the times should not be mistaken as underlying time-optimal solutions. The maximum number of loops was set to 3,000 for the concurrent update scheme and to 300,000 for the sequential update. The other numerical parameters were set as described in Section 5.9.1.

We recorded the achieved quality as a function of the running time and the number of the three matrix operations with the highest numerical demands, i.e., with a complexity of $\mathcal{O}(N^3)$: matrix multiplications, matrix eigendecompositions, and matrix exponentials. The last operation occured only in our implementation of the sequential-update algorithm, as described in Section 5.7.

In the following, we focus on results obtained with unconstrained optimisations whose initial control vectors had a mean value of 0 and a standard deviation of 1 (small amplitudes). These results can be found in Table E.1 in Appendix E and will be discussed below. Optimisations were also carried out using an initial control vector with a mean of value of 0 and a standard deviation of 10 (higher amplitudes), see Table E.2. Furthermore, we optimised the test suite with a constrained algorithm, using small amplitudes. These results are listed in Table E.3.

CHAPTER 5: BENCHMARKING A CONCURRENT-UPDATE OPTIMAL-CONTROL ALGORITHM

### 5.9.5.3 Results

Based on the data presented in Table E.1 (see Appendix E) and Figure 5.9.5, we note the following results:

First, in most of the problems, sequential- and concurrent-update algorithms reach similar final qualities. Out of the total of 23 test problems, the target quality of $10^{-4}$ is met in all cases with the exception of problems 5, 7, 10, 12, and 13. Only in problem 23, the sequential-update algorithm yields average final errors up to two orders of magnitude higher than the concurrent-update algorithm.

Second, the average running times differ substantially in many problems, with the concurrent-update algorithm being ahead. Only in problems 3, 4, 15, and 16, the final wall times are similar. Note that in all but the very easy problems 3, 4, and 16, the sequential algorithm needs a larger number of matrix multiplications and eigendecompositions. In particular, the sequential update scheme requires additional matrix exponentials in the forward propagation, which do not occur in concurrent-update.

In many problems (3, 5, 6, 8, 9, 11, 12, 14, 18, 19, 21, and 22), we observe a crossing point in the time course of the quality of the two algorithms. The sequential-update algorithm is overtaken by the concurrent-update scheme between a quality of 0.9 and 0.99 (see, e.g., Problem 21 in Figure 5.9.5). Therefore, dynamically changing from a sequential- to a concurrent-update scheme at a medium quality can be advantageous. This is exemplified in the optimisation shown in Figure 5.9.6: here, the sequential method is typically faster at the beginning of the optimisation, whereas the concurrent method overtakes at higher qualities near the end of the optimisation.

Moreover, with regard to dispersion of the final wall times required to achieve the target quality, the situation is fairly balanced: e.g., in problems 5 and 21 the sequential-update algorithm shows a larger relative standard deviation, while in problems 17 and 19 it is the concurrent-update algorithm.

We emphasise that the running times may strongly depend on the choice of initial conditions. Increasing mean value and standard deviation of the initial random control-amplitude vectors typically translates into longer wall times. This effect is more pronounced for the sequential- than for the concurrent-update algorithm. As a consequence, the performance differences between the two algorithms may increase and crossing or handover points may change. Results for higher initial pulse amplitudes with a higher standard deviation can be found in Table E.2 in Appendix E.

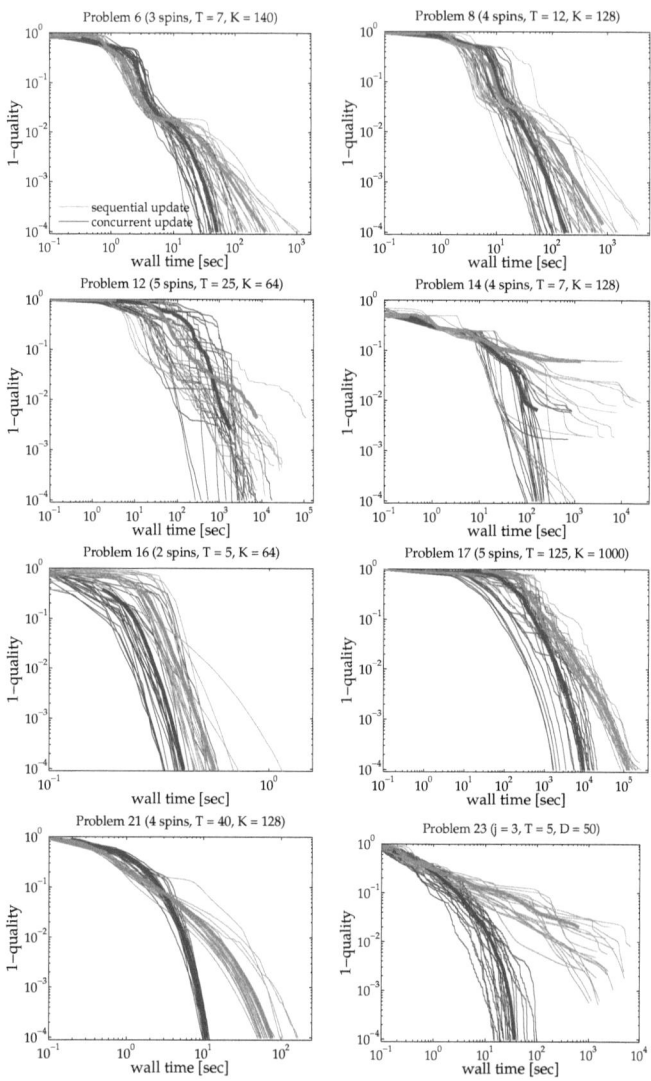

**Figure 5.9.5:** Doubly logarithmic plots showing the deviation from $q = 1$ as a function of the wall time for the concurrent- (blue) and the sequential-update algorithm (orange). Here, we present the results of only 8 out of 23 problems. The other plots can be found in Figures E.0.1 and E.0.2 in Appendix E.

CHAPTER 5: BENCHMARKING A CONCURRENT-UPDATE OPTIMAL-CONTROL ALGORITHM

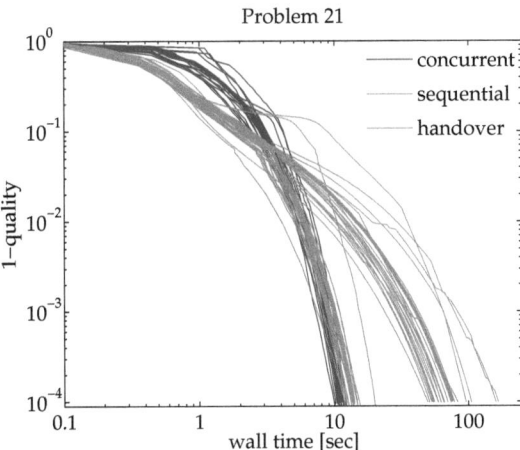

**Figure 5.9.6:** Example of a handover (green) from a sequential- (orange) to a concurrent-update (blue) algorithm. Based on the (unconstrained) optimisations of Problem 21 (see Figure 5.9.5, bottom left panel), the sequential-update algorithm is run up to a handover quality of 0.935, where the resulting pulse sequence is then used as the input to the concurrent-update algorithm for an optimisation up to the target quality of 0.9999.

Finally, as shown in Figure 5.9.7, the performance of the concurrent-update algorithm also differs between constrained and unconstrained optimisation, i.e., between the standard Matlab subroutines fmincon and fminunc (see Matlab documentation). In contrast, the sequential-update algorithm uses the same set of routines for both types of optimisations, where a basic cut-off method for respecting the constraints has almost no effect on the timings, as also illustrated by Figure 5.9.7.

### 5.9.5.4 Discussion

As expected from second-order versus first-order methods, at higher qualities (here typically $90 - 99\%$), the concurrent-update algorithm with BFGS overtakes the sequential-update algorithm using steepest descent. Therefore, changing from a sequential- to a concurrent-update algorithm at a medium quality can be advantageous. For reaching a quality of $1 - 10^{-4}$ in unitary gate synthesis, the concurrent-update algorithm is faster, in a number of instances even by more than one order of magnitude on average. Yet, at lower qualities, the computational speeds are similar and the sequential-update algorithm typically has a (small) advantage.

Chapter 5: Benchmarking a concurrent-update optimal-control algorithm

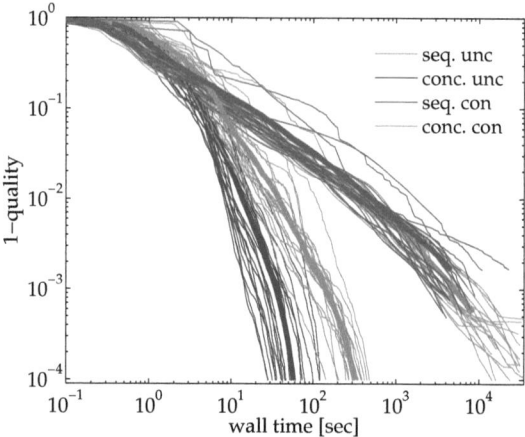

**Figure 5.9.7:** Doubly logarithmic plots showing the deviation from $q = 1$ as a function of the wall time for constrained and unconstrained concurrent-update (blue and green) and sequential-update (orange and pink) optimisations on model 3. The same initial conditions were used for the constrained and unconstrained runs. As expected, the sequential-update algorithm shows very similar wall times, whereas the concurrent-update algorithm is slowed down significantly when constraints are applied.

## 5.10 Conclusions

This chapter introduced a concurrent-update optimal control algorithm and presented (1) a set of benchmarks for components of this algorithm and (2) a comparison with a Krotov-type sequential-update algorithm.

For computing gradients (with respect to the control amplitudes), exact methods using the eigendecomposition in most cases proved superior to gradient approximations by finite differences, series expansions, or the standard method presented in [50]. If no exact gradient formula can be derived, however, finite-difference gradients provide a fast and easy-to-implement solution. The standard gradient approximation works well if Equation 5.5.5 is fulfilled - an issue that needs to be taken care of, as we saw in Section 5.9.3.

For computing the control vector update, the second-order BFGS method outperformed the first-order steepest ascent and the conjugate gradient methods. While using extra computational resources for approximating the Hessian, BFGS updated the control vector in a way that made it more efficient than the other two update methods.

In the last part of the chapter, we compared the performance of a gradient-based algorithm updating the time slots in the control vector in a sequential manner (Krotov-type) with an algorithm that featured a concurrent-update scheme. When it comes to implementing second-order update schemes, the different construction of the sequential- and concurrent-update algorithms translates into different performance: The concurrent-update algorithm matches particularly well with (quasi-)Newton methods, in particular a standard BFGS implementation. Currently, however, there seems to be no standard Newton-type second-order routine that would match with the sequential-update algorithm in a computationally fast and efficient way; finding such a routine is rather an open research problem. We employed efficient implementations, i.e., first-order gradient ascent for sequential update and a second-order concurrent update. As expected from second-order versus first-order methods, concurrent update overtakes sequential update at higher qualities (here typically $90 - 99\%$). For reaching a quality in unitary gate synthesis of $1 - 10^{-4}$, the concurrent-update algorithm is faster, in a number of instances even by more than one order of magnitude, on average. Yet at lower qualities, the computational speeds are similar and the sequential-update algorithm typically has a (small) advantage.

CHAPTER 6

# Conclusions and outlook

This thesis presents control-theoretical studies of dynamical quantum systems that can be used for quantum computation or simulation, and as quantum information resources, e.g., as quantum channels. In order to have useful applications in these fields, the systems must be either fully controllable or task-controllable, depending on the requirements of the desired application. For universal quantum computation, full controllability is required, whereas task controllability plays a crucial role in systems that are tailored to a specific task and therefore cannot or need not be made fully controllable.

Investigating controllability properties alone does not solve the problem of controlling a quantum system efficiently in practice, i.e., with experimental constraints. Solutions for this aspect are provided by another branch of control theory, namely optimal control. Here, the experimenter uses a numerical procedure based on Pontryagin's maximum principle to obtain control pulses steering the system in an optimal way. The optimisation generally takes into account experimental parameters and constraints while finding time-optimal or relaxation-optimal solutions. In our case, this procedure is an iterative scheme known as the GRAPE algorithm.

Chapter 3 introduces quantum dynamical control systems and investigates their symmetry restraints to establish sufficient conditions for full and task controllability. Based on a unified Lie-algebraic framework, the interplay of drift and control Hamiltonians is analysed for closed qubit systems with an arbitrary coupling graph. In particular, for a given system, the dynamical system Lie algebra explicitly yields the reachable set which defines the feasible tasks on that system. We find that symmetries eliminate full controllability and thus can be harmful when designing a quantum system. For any non-symmetric qubit system, we provide three design rules that ensure full controllability for couplings of the follow-

## CHAPTER 6: CONCLUSIONS AND OUTLOOK

ing types: Ising-$ZZ$, Heisenberg-$XXX$, Heisenberg-$XXZ$, Heisenberg-$XYZ$, Heisenberg-$XX$, and Heisenberg-$XY$. Reference [96] translates the restrictions on controllability into restrictions on *observability*. The classical duality of these two concepts does not hold 1:1 in the quantum domain. Of further interest are conditions that fill the gap between lack of symmetry and full controllability in qubit systems. These issues are addressed in [67].

Another application of Lie-algebraic methods is presented in Chapter 4: a gradient flow on the unitary group provides a numerical method to optimise channel capacities. These capacities have long been assumed to be additive, with one counterexample given by Werner and Holevo [68]. Our goal is to find more counterexamples by testing a large set of channels for their additivity properties. While we successfully optimise capacities of standard and random unitary channels, this search is found to be unsuccessful eventually, mainly because the search space of possible channels is too large for a numerical method. The tested channels are of relatively small dimensions ($2 < d < 6$), in particular compared to the succeeding disproofs of the additivity conjecture in [75] and [76] where high-dimensional unitary channels were studied. The conjecture is now disproven, but the Werner-Holevo channel remains the only explicit counterexample, especially in small dimensions. Further research may focus on finding concrete examples of this type.

Chapter 5 discusses and benchmarks an iterative optimal control algorithm with a concurrent control-update scheme. After introducing the optimal control framework, the algorithm is presented in a modular form. Different methods for computing gradients and for updating the control vector are described and compared numerically using a set of test systems. It is shown that the performance of the algorithm depends on the interplay of all its modules, which must be kept in mind when one wants to exchange a module. In our tests, the best performance is achieved by an exact gradient routine combined with a second-order update scheme based on BFGS. In particular, the BFGS method is found to be superior to a simple steepest ascent method and to conjugate gradients. Using exact gradients, if possible, is generally advantageous to relying on approximated gradients. When no exact gradient formula is known, finite-difference gradients provide a useful and fast alternative. In comparison to an established Krotov-type sequential-update algorithm, the concurrent-update algorithm is faster in reaching high qualities of $1 - 10^{-4}$, partly because there is no known second-order method for the sequential-update algorithm that is matched to it as well as BFGS is to the concurrent-update algorithm. As sequential and concurrent updates represent two extrema of a spectrum, one can think of hybrid versions that update a subset of available time slots. A modular structure allows an easy combination of different approaches, as would be needed for hybrids. Further development can include an advanced optimisation of (higher-order) update strategies, in particular for a sequential-update scheme, and concepts for parallelising optimal control algorithms for their use on high performance clusters.

# Appendix A

# The superoperator formalism 101

Here, we would like to outline the basic notions of the superoperator formalism; a full treatment of the can be found in [57].

A superoperator is a linear operator that acts on a vector space of linear operators. A superoperator can be represented as a matrix if the operator it is acting on are represented as vectors. This can be done in the following way.

Consider a 'vec' operation that maps a matrix $\in \text{Mat}_N(\mathbb{C})$ onto a vector $v \in \mathbb{C}^{N^2}$: $\text{Mat}_N(\mathbb{C}) \to \mathbb{C}^{N^2}$. This vector can be computed by stacking up the columns of $M$ on top of each other:

$$M \to \begin{bmatrix} \text{vec } M_1 \\ \vdots \\ \text{vec } M_N \end{bmatrix}.$$

Here, vec $M_k$ again denotes the $k$-th column vector of $M$. Another notation for vec $M$ is $|M\rangle$. As an example, consider the conjugation of $X \in \text{Mat}_N(\mathbb{C})$ with $M \in \mathbf{GL}(N)$, i.e., $MXM^{-1}$. Using the vec operation, this conjugation can be rewritten as

$$MXM^{-1} \to \left((M^{-1})^t \otimes M\right)\text{vec}(X).$$

Shorter variations of this expression are $\text{Ad}_M\text{vec}(X)$ or $\hat{M}|X\rangle$. In contrast, the expression $\text{ad}_M(\cdot)$ denotes the commutator with $M$, see Section 3.2. In the superoperator formalism, the computation of traces of matrices $X$ and $Y$ becomes a computation of an inner product:

$$\text{tr}(X^\dagger Y) \to \langle X|Y\rangle = \left(\text{vec}(X^\dagger)\right)^t \left(\text{vec}(Y)\right).$$

# Appendix B

# Proof of Lemma 4

The following proof is based on [97].

*Proof.* Let $A$ and $B$ be Hermitian matrices. Let $\{|\xi_\alpha\rangle\}$ be the set of orthonormal eigenvectors to the eigenvalues $\{\xi_\alpha\}$ of $A$. Then, the following equalities hold:

$$\begin{aligned}
\langle \xi_\mu | \frac{d}{dx} e^{A+xB} | \xi_\nu \rangle \bigg|_{x=0} &= \langle \xi_\mu | \frac{d}{dx} \sum_{n=0}^{\infty} \frac{1}{n!} (A+xB)^n | \xi_\nu \rangle \bigg|_{x=0} \\
&= \langle \xi_\mu | \sum_{n=0}^{\infty} \frac{1}{n!} \sum_{q=1}^{n} (A+xB)^{q-1} B (A+xB)^{n-q} | \xi_\nu \rangle \bigg|_{x=0} \\
&= \langle \xi_\mu | \sum_{n=0}^{\infty} \frac{1}{n!} \sum_{q=1}^{n} A^{q-1} B A^{n-q} | \xi_\nu \rangle \\
&= \sum_{n=0}^{\infty} \frac{1}{n!} \sum_{q=1}^{n} \xi_\mu^{q-1} \langle \xi_\mu | B | \xi_\nu \rangle \xi_\nu^{n-q}. \qquad (B.0.1)
\end{aligned}$$

If $\xi_\mu = \xi_\nu$, one can immediately see that

$$\langle \xi_\mu | \frac{d}{dx} e^{A+xB} | \xi_\nu \rangle \bigg|_{x=0} = \langle \xi_\mu | B | \xi_\nu \rangle e^{\xi_\mu}. \qquad (B.0.2)$$

If $\xi_\mu \neq \xi_\nu$, we obtain:

$$\langle \xi_\mu | \frac{d}{dx} e^{A+xB} | \xi_\nu \rangle \bigg|_{x=0} = \langle \xi_\mu | B | \xi_\nu \rangle \sum_{n=0}^{\infty} \frac{1}{n!} \sum_{q=1}^{n} \xi_\mu^{q-1} \xi_\nu^{n-q}$$

# APPENDIX B: PROOF OF LEMMA 4

$$\begin{aligned}
&= \langle \xi_\mu | B | \xi_\nu \rangle \sum_{n=0}^{\infty} \frac{1}{n!} \xi_\nu^{n-1} \sum_{q=1}^{n} \left( \frac{\xi_\mu}{\xi_\nu} \right)^{q-1} \\
&= \langle \xi_\mu | B | \xi_\nu \rangle \sum_{n=0}^{\infty} \frac{1}{n!} \xi_\nu^{n-1} \frac{(\xi_\mu / \xi_\nu)^n - 1}{(\xi_\mu / \xi_\nu) - 1} \\
&= \langle \xi_\mu | B | \xi_\nu \rangle \sum_{n=0}^{\infty} \frac{1}{n!} \frac{\xi_\mu^n - \xi_\nu^n}{\xi_\mu - \xi_\nu} \\
&= \langle \xi_\mu | B | \xi_\nu | \rangle \frac{e^{\xi_\mu} - e^{\xi_\nu}}{\xi_\mu - \xi_\nu}.
\end{aligned}$$

$\square$

An analogous result is found for the skew-Hermitian matrices $iA$ and $iB$. This covers the case discussed in Section 5.5.1.2, where $A := -i\Delta t_k H_{tot}(t_k)$ and $B := -i\Delta t_k H_m$.

Note that the proof requires the eigenvectors for different eigenvalues to be orthogonal. Thus, it only holds for (skew-)Hermitian matrices describing closed quantum systems. A generalisation to open quantum systems cannot be made in a simple way.

APPENDIX C

# Comparison of gradient methods

In addition to the results shown for model 1 in Section 5.9.3, here the results for the optimisation of model 2 will be presented. We chose final times of $T = 14$ sec (unitary optimisation) and $T = 1$ sec (state-to-state optimisation). Again, we did not aim for minimal times. The target was a cluster state or the unitary gate generating this cluster state.

Table C.1 lists the values of $P = \text{mean}\left(\Delta t \cdot ||H_{tot}(t_k)||_2\right)$ used for these tests.

*Unitary optimisation*: values of $P = \text{mean}\left(\Delta t \cdot ||H_{tot}||_2\right)$, $(T = 14 \text{ s})$.

| s \ K | 50 | 80 | 100 | 200 |
|---|---|---|---|---|
| 0.1 | 0.8419 | 0.5262 | 0.4209 | 0.2105 |
| 10 | 6.9670 | 4.4204 | 3.5081 | 1.7592 |

*State-to-state optimisation*: values of $P = \text{mean}\left(\Delta t \cdot ||H_{tot}||_2\right)$, $(T = 1 \text{ s})$.

| s \ K | 20 | 50 | 80 | 120 |
|---|---|---|---|---|
| 1 | 0.1855 | 0.0741 | 0.0463 | 0.0309 |
| 20 | 2.4876 | 1.0009 | 0.6257 | 0.4190 |

**Table C.1:** Model 2: Values of $P$ for the different numbers of time slots $K$ and initial amplitudes which are represented by the standard deviation of the initial pulse sequence (denoted by $s$). The top table shows the values for the unitary optimisations, the bottom one shows the values for the optimisations of state-to-state transfers.

The results obtained for model 2 are similar to those for model 1, see Figures C.0.1 and C.0.2.

APPENDIX C: COMPARISON OF GRADIENT METHODS

The exact gradient method (blue lines) performs best in all cases (a)-(d). The standard approximation (pink lines) breaks down and fails to reach the goal quality $q_{goal}$ in cases (a) and (b). In the other cases, most runs using the standard approximation succed in achieving $q_{goal}$. The speed of this method is then comparable to the speed of the exact gradient method.

In comparison to these two methods, the finite-difference method (green) and the Hausdorff series (orange) are slower while achieving $q_{goal}$ in all cases.

APPENDIX C: COMPARISON OF GRADIENT METHODS

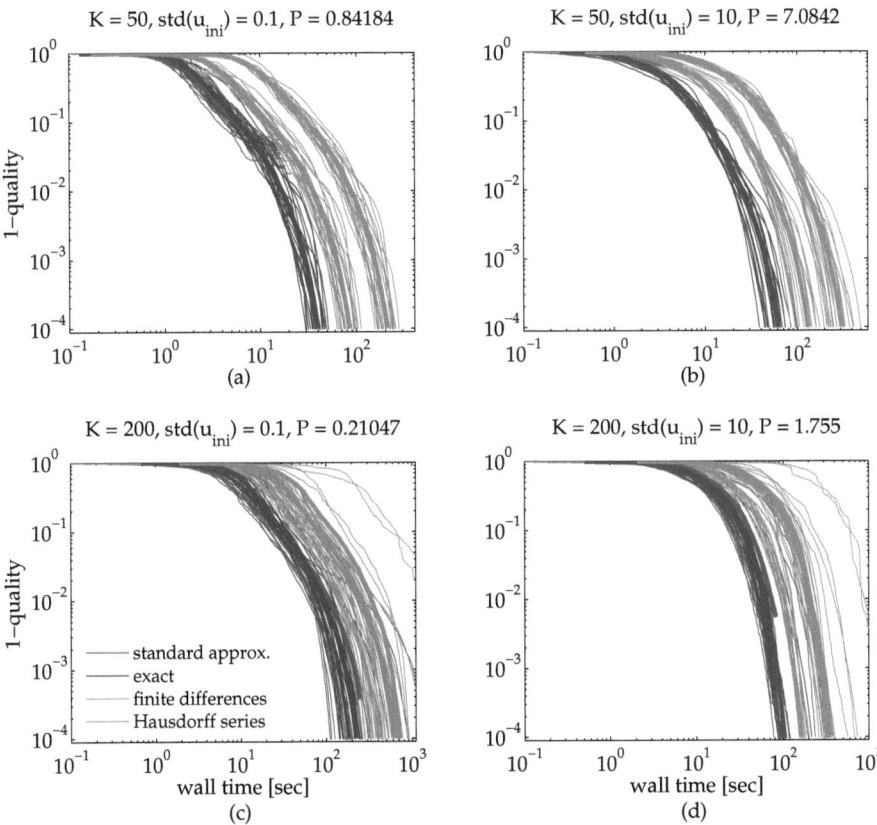

**Figure C.0.1:** Performance of the four gradient methods discussed in subsection 5.5.1 for the optimisation of a unitary gate for model 2. The exact gradient method (blue line) performs best in all cases. We observe the breakdown of the standard approximation (pink) for cases (a) and (b). For the higher digitisation of $M = 128$, the finite-difference gradient method (green) and the Hausdorff series (orange) perform significantly slower than the other two methods. The thick lines represent average values. Note that the plot is doubly logarithmic.

In Fig. C.0.2, we observe the breakdown of the standard approximation (green and orange lines) for values of $P$ that are of the order of $10^{-1}$, with small differences between unitary optimisations and state-to-state optimisations. The exact gradient method (dashed blue lines) yields excellent mean final qualities for all values of $P$.

APPENDIX C: COMPARISON OF GRADIENT METHODS

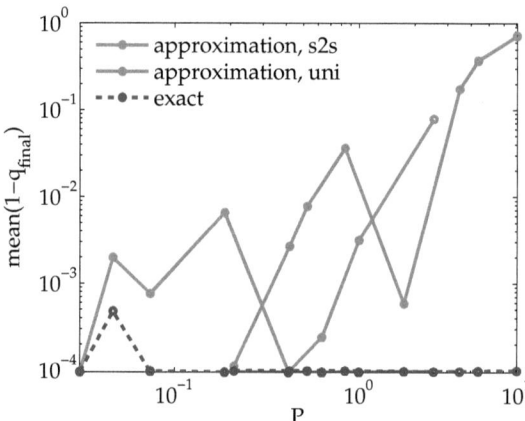

**Figure C.0.2:** Doubly logarithmic plot showing the mean final qualities achieved by the exact gradient method (dashed blue lines) and the standard approximation on model 2. The green line represents optimisations of state-to-state transfers (cluster state generation), the orange line represents optimisations for the unitary gate that represents a cluster state generation. The breakdown of the standard approximation for high values of $P$ is observed.

APPENDIX D

# The interior-point algorithm

The interior-point method used for the constrained optimisation problem in Sec. 5.9 was developed [98, 99] to solve large-scale constrained minimisation problems of the form

$$\min_x f(x)$$
$$\text{subject to } h(x) = 0,$$
$$\text{and } g(x) \leq 0, \tag{D.0.1}$$

where $f : \mathbb{R}^n \to \mathbb{R}$, $h : \mathbb{R}^n \to \mathbb{R}^n$, and $g : \mathbb{R}^n \to \mathbb{R}^m$ are smooth functions. It is interior in the sense that it starts searching with a feasible point, i.e., a point that satisfies the constraints, as opposed to exterior methods which start with points outside the feasible area of the search space. The general idea of this interior-point approach is to solve a sequence of approximate subproblems that are parametrised by a parameter $\mu > 0$. These subproblems are of the form

$$\min_{x,s} f(x) - \mu \sum_{i=1}^{m} \ln s_i$$
$$\text{subject to } h(x) = 0, \tag{D.0.2}$$
$$\text{and } g(x) + s = 0 \tag{D.0.3}$$

The added logarithmic term is called a *barrier function* with the *barrier parameter* $\mu$ that has typically a large initial value. The barrier function is, in a sense, the opposite of a penalty function for exterior methods. The slack variables $s_i$ are assumed to be positive, and their number equals the number of inequality constraints $g$. When $\mu$ converges to zero, the sequence of solutions of (D.0.2) should converge to the minimum of the original problem

## Appendix D: The Interior-Point Algorithm

(D.0.1). Hence, we have to solve a sequence of equality-constrained problems, which is easier than solving the inequality-constrained problem (D.0.1). As described in [98], the subproblems are solved by a combination of sequential linear programming and trust-region techniques. At each iteration, the algorithm uses either a direct step (also called Newton step) or, if the direct step cannot be taken, a conjugate gradient step. One particular advantage is that second-order information can be used directly. Thus, the algorithm is well suited to be combined with the Broyden-Fletcher-Goldfarb-Shanno method for efficiently computing an approximation of the Hessian matrix.

APPENDIX E

# Additional numerical results

In this appendix, we present additional numerical data that was collected for the comparison of the sequential- and concurrent-update algorithms, as described in Sec. 5.9.5.

Figs. E.0.1 and E.0.2 depict the achieved quality as a function of the running time for all problems mentioned in Sec. 5.9.5, except those problems already included in Fig. 5.9.5 in that section.

The Tabs. E.1, E.2, and E.3 list minimal, maximal, and mean values of quantities that were measured during the optimisations using the sequential- and the concurrent-update algorithms. In particular, the wall times, the achieved qualities, and the number of expensive matrix operations can found in these tables.

APPENDIX E: ADDITIONAL NUMERICAL RESULTS

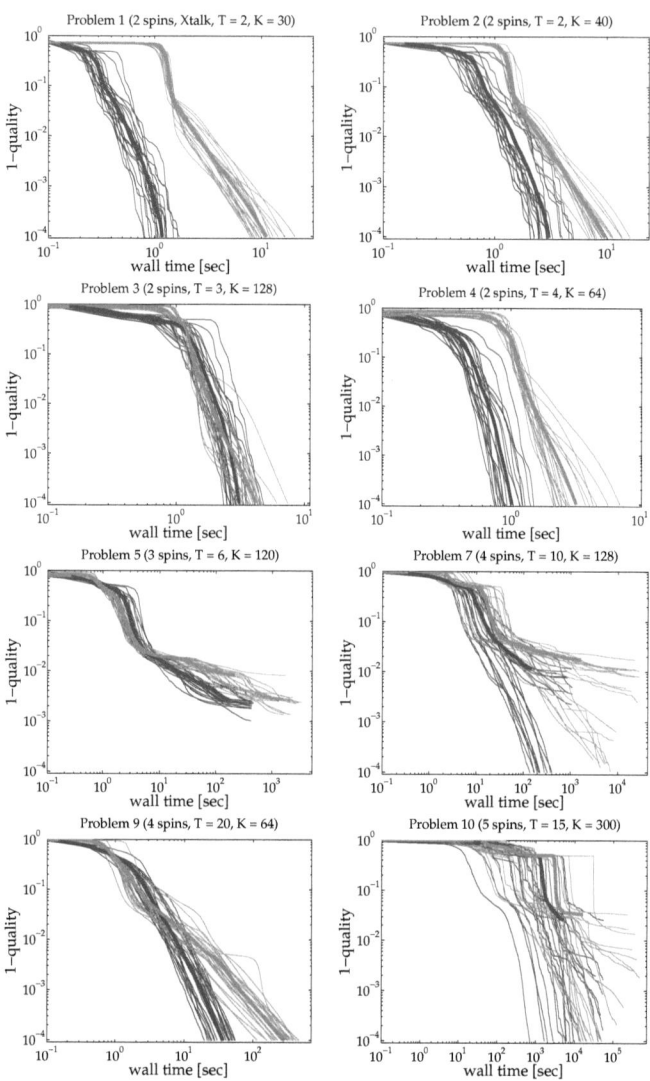

**Figure E.0.1:** Doubly logarithmic plots showing the deviation from $q = 1$ as a function of the wall time for the concurrent- (blue) and the sequential-update algorithm (orange).

APPENDIX E: ADDITIONAL NUMERICAL RESULTS

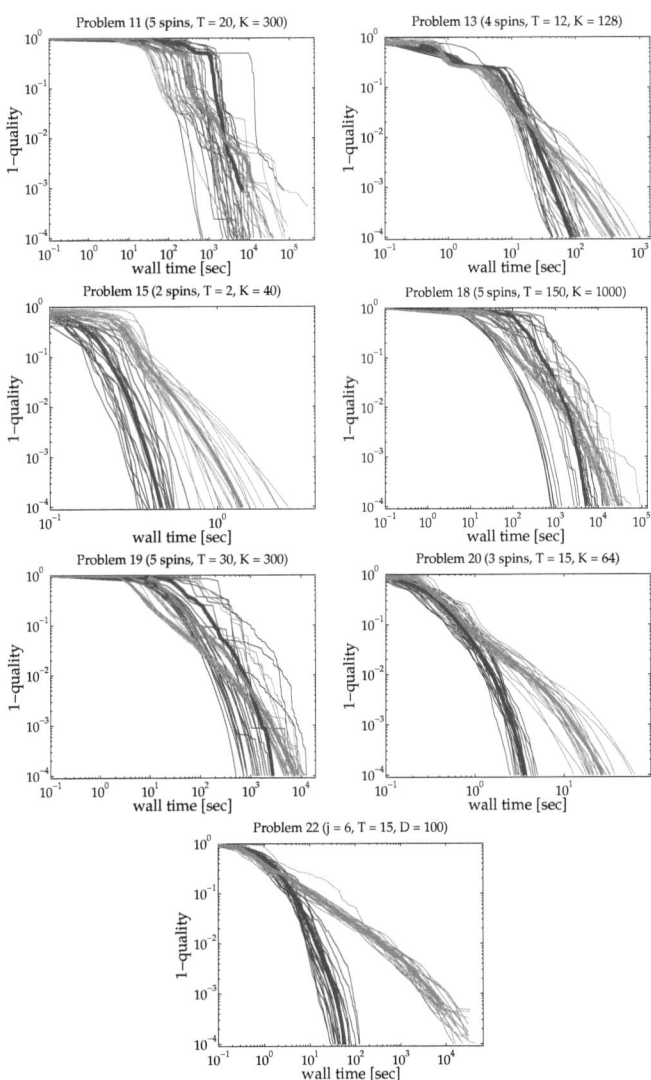

**Figure E.0.2:** Doubly logarithmic plots showing the deviation from $q = 1$ as a function of the wall time for the concurrent- (blue) and the sequential-update algorithm (orange).

APPENDIX E: ADDITIONAL NUMERICAL RESULTS

**Table E.1:** Overview on test results obtained from 20 *unconstrained optimisations* (fminunc in Matlab) for each problem listed in Tab. 5.2 using the sequential or the concurrent-update algorithm. *Small initial pulse amplitudes* ((mean$(u_{ini}) = 0$, std$(u_{ini}) = 1$)) were used. Mean values are in bold face. Note that the concurrent algorithm needs no extra matrix exponentials (see text).

| Problem | Algorithm | Final Quality (mean/min/max) | Wall Time [min] (mean/min/max) | #Eigendecs/1000 (mean/min/max) | #Matrix Mults/1000 (mean/min/max) | #Matrix Exps/1000 (mean/min/max) |
|---|---|---|---|---|---|---|
| 1 | conc. | **0.9999**/0.9999/1.0000 | **0.02**/0.01/0.03 | **2.02**/1.35/2.94 | **38**/25/56 | —none— |
|   | seq.  | **0.9999**/0.9999/0.9999 | **0.19**/0.13/0.34 | **6.29**/4.36/11.68 | **88**/61/163 | **6.29**/4.36/11.68 |
| 2 | conc. | **0.9999**/0.9999/1.0000 | **0.05**/0.03/0.08 | **2.68**/1.76/4.44 | **50**/33/84 | — |
|   | seq.  | **0.9999**/0.9999/0.9999 | **0.16**/0.11/0.27 | **5.43**/3.80/9.04 | **76**/53/126 | **5.43**/3.80/9.04 |
| 3 | conc. | **0.9999**/0.9999/1.0000 | **0.05**/0.04/0.08 | **4.61**/3.46/7.04 | **85**/63/132 | — |
|   | seq.  | **0.9999**/0.9999/0.9999 | **0.07**/0.05/0.12 | **2.29**/1.56/4.12 | **32**/22/57 | **2.29**/1.56/4.12 |
| 4 | conc. | **0.9999**/0.9999/1.0000 | **0.02**/0.01/0.02 | **1.70**/1.28/2.43 | **31**/23/45 | — |
|   | seq.  | **0.9999**/0.9999/0.9999 | **0.05**/0.03/0.11 | **1.72**/1.08/3.84 | **24**/15/54 | **1.72**/1.08/3.84 |
| 5 | conc. | **0.9978**/0.9973/0.9990 | **7.19**/5.84/7.86 | **362**/310/367 | **9774**/8364/9917 | — |
|   | seq.  | **0.9973**/0.9918/0.9986 | **34**/22/58 | **1976**/1320/3292 | **35542**/23729/59209 | **1976**/1320/3292 |
| 6 | conc. | **0.9999**/0.9999/0.9999 | **0.85**/0.34/2.21 | **35**/17/76 | **954**/450/2050 | — |
|   | seq.  | **0.9999**/0.9999/0.9999 | **5.14**/1.14/18.72 | **310**/68/1143 | **5574**/1216/20554 | **310**/68/1143 |
| 7 | conc. | **0.9970**/0.9886/0.9999 | **9.42**/2.32/18.48 | **229**/63/391 | **8028**/2210/13679 | — |
|   | seq.  | **0.9945**/0.9825/0.9999 | **242**/50/491 | **3975**/1242/7753 | **87385**/27313/170455 | **3975**/1242/7753 |
| 8 | conc. | **0.9999**/0.9999/0.9999 | **2.90**/0.83/10.39 | **72**/21/275 | **2530**/735/9627 | — |
|   | seq.  | **0.9999**/0.9999/0.9999 | **16.11**/2.36/65.72 | **500**/89/2223 | **11002**/1953/48876 | **500**/89/2223 |
| 9 | conc. | **0.9999**/0.9999/0.9999 | **0.61**/0.33/0.92 | **20**/11/30 | **685**/381/1052 | — |
|   | seq.  | **0.9999**/0.9999/0.9999 | **4.83**/1.91/7.81 | **161**/63/259 | **3536**/1375/5696 | **161**/63/259 |
| 10 | conc. | **0.9982**/0.9740/0.9999 | **376**/12/918 | **435**/82/917 | **18694**/3510/39442 | — |
|    | seq.  | **0.9959**/0.9661/0.9999 | **2591**/244/8458 | **13123**/1312/40136 | **341107**/34116/1043279 | **13123**/1312/40136 |
| 11 | conc. | **0.9999**/0.9991/0.9999 | **148**/11/1236 | **189**/71/919 | **8114**/3045/39519 | — |
|    | seq.  | **0.9998**/0.9988/0.9999 | **786**/72/4817 | **4041**/427/17767 | **103031**/11097/461821 | **4041**/427/17767 |
| 12 | conc. | **0.9996**/0.9974/0.9999 | **62.22**/4.48/286.60 | **89**/22/192 | **3818**/942/8276 | — |
|    | seq.  | **0.9994**/0.9956/0.9999 | **284**/56/1842 | **987**/245/4563 | **25637**/6360/118491 | **987**/245/4563 |
| 13 | conc. | **0.9999**/0.9999/0.9999 | **1.45**/0.70/2.62 | **35**/19/60 | **1235**/677/2089 | — |
|    | seq.  | **0.9999**/0.9999/0.9999 | **6.47**/1.76/16.06 | **219**/59/547 | **4813**/1292/12033 | **219**/59/547 |
| 14 | conc. | **0.9989**/0.9936/0.9999 | **5.20**/1.52/14.89 | **138**/41/390 | **4833**/1434/13634 | — |
|    | seq.  | **0.9759**/0.9373/0.9999 | **129.00**/6.39/439.92 | **4174**/215/15103 | **91773**/4719/332029 | **4174**/215/15103 |
| 15 | conc. | **0.9999**/0.9999/1.0000 | **0.01**/0.00/0.01 | **0.90**/0.64/1.24 | **9.57**/6.67/13.30 | — |
|    | seq.  | **0.9999**/0.9999/1.0000 | **0.02**/0.01/0.04 | **1.76**/0.72/3.28 | **17.53**/7.16/32.64 | **1.76**/0.72/3.28 |
| 16 | conc. | **0.9999**/0.9999/1.0000 | **0.01**/0.00/0.01 | **0.80**/0.70/1.28 | **8.19**/7.13/13.48 | — |
|    | seq.  | **0.9999**/0.9999/1.0000 | **0.01**/0.01/0.02 | **0.67**/0.51/1.54 | **6.64**/5.10/15.31 | **0.67**/0.51/1.54 |
| 17 | conc. | **0.9999**/0.9999/0.9999 | **160**/27/357 | **684**/616/773 | **7516**/6767/8495 | — |
|    | seq.  | **0.9999**/0.9999/0.9999 | **2582**/1411/4638 | **16577**/11490/27082 | **165733**/114877/270766 | **16577**/11490/27082 |
| 18 | conc. | **0.9999**/0.9999/0.9999 | **88**/13/220 | **394**/286/620 | **4330**/3137/6811 | — |
|    | seq.  | **0.9999**/0.9999/0.9999 | **492**/247/1520 | **2954**/2434/3985 | **29535**/24335/39842 | **2954**/2434/3985 |
| 19 | conc. | **0.9999**/0.9999/0.9999 | **45.85**/8.49/213.95 | **170**/103/264 | **3896**/2354/6060 | — |
|    | seq.  | **0.9999**/0.9999/0.9999 | **128**/76/217 | **1124**/809/1490 | **17978**/12945/23822 | **1124**/809/1490 |
| 20 | conc. | **0.9999**/0.9999/0.9999 | **0.06**/0.04/0.08 | **6.92**/4.80/9.47 | **76**/52/104 | — |
|    | seq.  | **0.9999**/0.9999/0.9999 | **0.45**/0.21/1.02 | **26**/15/43 | **258**/148/431 | **26**/15/43 |
| 21 | conc. | **0.9999**/0.9999/0.9999 | **0.18**/0.16/0.20 | **8.56**/7.81/9.60 | **161**/146/180 | — |
|    | seq.  | **0.9999**/0.9999/0.9999 | **1.26**/0.82/2.76 | **39**/29/57 | **551**/410/804 | **39**/29/57 |
| 22 | conc. | **0.9999**/0.9947/2.02 | **0.96**/0.47/2.02 | **68**/42/105 | **750**/459/1154 | — |
|    | seq.  | **0.9998**/0.9994/0.9999 | **407**/112/732 | **21692**/6473/30000 | **216483**/64599/299399 | **21692**/6473/30000 |
| 23 | conc. | **0.9999**/0.9999/0.9999 | **0.60**/0.24/1.64 | **53**/25/141 | **588**/279/1559 | — |
|    | seq.  | **0.9951**/0.9797/0.9995 | **39.03**/9.39/111.74 | **2992**/744/7163 | **29796**/7408/71343 | **2992**/744/7163 |

APPENDIX E: ADDITIONAL NUMERICAL RESULTS

**Table E.2:** Overview on test results obtained from 20 *unconstrained optimisations* (`fminunc` in Matlab) for each problem listed in Tab. 5.2 using sequential or concurrent-update. *Higher initial pulse amplitudes* (mean$(u_{ini}) = 0$, std$(u_{ini}) = 10$) than in Tab. E.1 were used. Mean values are in bold face. Note that the concurrent algorithm needs no extra matrix exponentials (see text).

| Problem | Algorithm | Final Quality (mean/min/max) | Wall Time [min] (mean/min/max) | #Eigendecs/1000 (mean/min/max) | #Matrix Mults/1000 (mean/min/max) | #Matrix Exps/1000 (mean/min/max) |
|---|---|---|---|---|---|---|
| 1 | conc. | **0.9999**/0.9999/0.9999 | **0.04**/0.02/0.06 | **4.25**/2.61/6.60 | **80**/49/125 | —none— |
|   | seq.  | **0.9999**/0.9998/0.9999 | **1.54**/0.43/3.78 | **118**/33/289 | **1645**/459/4023 | **118**/33/289 |
| 2 | conc. | **0.9999**/0.9999/0.9999 | **0.05**/0.02/0.08 | **5.39**/2.76/8.56 | **102**/52/162 | — |
|   | seq.  | **0.9999**/0.9999/0.9999 | **1.38**/0.42/3.28 | **109**/33/261 | **1520**/464/3635 | **109**/33/261 |
| 3 | conc. | **0.9999**/0.9999/1.0000 | **0.07**/0.05/0.10 | **6.06**/4.35/8.45 | **113**/80/158 | — |
|   | seq.  | **0.9999**/0.9999/0.9999 | **0.22**/0.12/0.49 | **17.29**/9.73/38.53 | **242**/136/539 | **17.29**/9.73/38.53 |
| 4 | conc. | **0.9999**/0.9999/1.0000 | **0.02**/0.01/0.03 | **2.50**/1.60/3.39 | **46**/29/63 | — |
|   | seq.  | **0.9999**/0.9999/0.9999 | **0.18**/0.05/0.54 | **13.79**/4.22/42.18 | **193**/59/589 | **13.79**/4.22/42.18 |
| 5 | conc. | **0.9976**/0.9959/0.9986 | **6.97**/6.24/7.32 | **364**/362/370 | **9839**/9784/9995 | — |
|   | seq.  | **0.9969**/0.9952/0.9983 | **73**/51/105 | **4246**/2954/6021 | **76349**/53121/108271 | **4246**/2954/6021 |
| 6 | conc. | **0.9999**/0.9999/0.9999 | **2.23**/1.36/3.82 | **105**/60/180 | **2842**/1623/4860 | — |
|   | seq.  | **0.9999**/0.9999/0.9999 | **16**/11/32 | **935**/614/1866 | **16823**/11049/33567 | **935**/614/1866 |
| 7 | conc. | **0.9893**/0.9366/0.9999 | **15**/14/16 | **386**/385/387 | **13499**/13469/13563 | — |
|   | seq.  | **0.9928**/0.9444/0.9993 | **325**/129/730 | **8053**/4190/16630 | **177047**/92125/365595 | **8053**/4190/16630 |
| 8 | conc. | **0.9998**/0.9984/0.9999 | **9.86**/4.27/15.02 | **257**/110/386 | **9000**/3832/13495 | — |
|   | seq.  | **0.9990**/0.9851/0.9999 | **158**/43/645 | **3511**/1400/13269 | **77189**/30788/291716 | **3511**/1400/13269 |
| 9 | conc. | **0.9999**/0.9999/0.9999 | **2.81**/1.80/4.62 | **87**/57/142 | **3056**/2007/4982 | — |
|   | seq.  | **0.9996**/0.9995/0.9998 | **41**/22/104 | **1057**/693/2033 | **23223**/15223/44653 | **1057**/693/2033 |
| 10 | conc. | **0.9978**/0.9834/0.9999 | **638**/91/1566 | **798**/402/944 | **34315**/17276/40590 | — |
|    | seq.  | **0.9995**/0.9974/0.9999 | **3213**/529/8282 | **16045**/6914/33844 | **417076**/179721/879708 | **16045**/6914/33844 |
| 11 | conc. | **0.9999**/0.9998/0.9999 | **449**/37/1165 | **613**/335/906 | **26342**/14386/38939 | — |
|    | seq.  | **0.9999**/0.9998/0.9999 | **1557**/863/2935 | **8408**/4895/14865 | **218564**/127232/386383 | **8408**/4895/14865 |
| 12 | conc. | **0.9984**/0.9948/0.9999 | **197**/16/416 | **196**/192/202 | **8426**/8273/8678 | — |
|    | seq.  | **0.9974**/0.9911/0.9990 | **883**/255/2060 | **4320**/1994/9522 | **112192**/51780/247266 | **4320**/1994/9522 |
| 13 | conc. | **0.9999**/0.9999/0.9999 | **1.48**/1.09/1.94 | **40**/33/48 | **1405**/1152/1676 | — |
|    | seq.  | **0.9999**/0.9999/0.9999 | **5.25**/3.90/6.50 | **166**/126/207 | **3654**/2772/4350 | **166**/126/207 |
| 14 | conc. | **0.9999**/0.9999/0.9999 | **2.65**/1.74/4.43 | **64**/47/107 | **2247**/1636/3729 | — |
|    | seq.  | **0.9999**/0.9999/0.9999 | **16.31**/9.63/31.65 | **520**/310/1040 | **11427**/6804/22872 | **520**/310/1040 |
| 15 | conc. | **0.9999**/0.9999/1.0000 | **0.00**/0.00/0.01 | **0.55**/0.40/0.76 | **5.70**/4.02/8.00 | — |
|    | seq.  | **0.9999**/0.9999/1.0000 | **0.01**/0.01/0.02 | **0.75**/0.52/1.60 | **7.44**/5.17/15.92 | **0.75**/0.52/1.60 |
| 16 | conc. | **0.9999**/0.9999/1.0000 | **0.00**/0.00/0.01 | **0.76**/0.58/1.22 | **7.73**/5.71/12.77 | — |
|    | seq.  | **0.9999**/0.9999/1.0000 | **0.01**/0.01/0.02 | **0.58**/0.45/1.15 | **5.77**/4.47/11.48 | **0.58**/0.45/1.15 |
| 17 | conc. | **0.9999**/0.9999/0.9999 | **162**/28/320 | **616**/536/750 | **6768**/5887/8242 | — |
|    | seq.  | **0.9999**/0.9999/0.9999 | **1346**/763/2603 | **9238**/7502/13230 | **92361**/75005/132274 | **9238**/7502/13230 |
| 18 | conc. | **0.9999**/0.9999/0.9999 | **118**/24/309 | **522**/400/652 | **5736**/4391/7163 | — |
|    | seq.  | **0.9999**/0.9999/0.9999 | **897**/428/1152 | **6788**/5799/8207 | **67863**/57978/82054 | **6788**/5799/8207 |
| 19 | conc. | **0.9999**/0.9999/0.9999 | **41.77**/5.48/120.52 | **65**/58/71 | **1481**/1332/1636 | — |
|    | seq.  | **0.9999**/0.9999/0.9999 | **59**/32/89 | **460**/398/547 | **7354**/6362/8747 | **460**/398/547 |
| 20 | conc. | **0.9998**/0.9991/0.9999 | **1.91**/0.51/6.98 | **139**/47/202 | **1529**/517/2225 | — |
|    | seq.  | **0.9980**/0.9879/0.9996 | **64**/29/103 | **3508**/2098/6647 | **34972**/20910/66265 | **3508**/2098/6647 |
| 21 | conc. | **0.9999**/0.9999/0.9999 | **1.50**/1.18/2.07 | **70**/53/96 | **1328**/1005/1818 | — |
|    | seq.  | **0.9998**/0.9998/0.9999 | **118**/84/164 | **4269**/2860/5791 | **59697**/39992/80980 | **4269**/2860/5791 |
| 22 | conc. | **0.9999**/0.9999/0.9999 | **0.58**/0.32/0.90 | **51**/29/74 | **563**/317/820 | — |
|    | seq.  | **0.9995**/0.9982/0.9998 | **81**/35/137 | **4128**/1981/6114 | **41194**/19771/61017 | **4128**/1981/6114 |
| 23 | conc. | **0.9999**/0.9999/0.9999 | **0.06**/0.05/0.07 | **7.58**/6.20/9.85 | **83**/68/108 | — |
|    | seq.  | **0.9999**/0.9999/0.9999 | **1.20**/0.59/2.24 | **93**/46/171 | **925**/458/1702 | **93**/46/171 |

# Appendix E: Additional numerical results

**Table E.3:** Overview on test results obtained from 20 *constrained optimisations* (fmincon in Matlab) for each problem listed in Tab. 5.2 using the sequential or the concurrent-update algorithm. *Small initial pulse amplitudes* (mean$(u_{ini}) = 0$, std$(u_{ini}) = 1$) were used. Mean values are in bold face. Note that the concurrent algorithm needs no extra matrix exponentials (see text).

| Problem | Algorithm | Final Quality (**mean**/min/max) | Wall Time [min] (**mean**/min/max) | #Eigendecs/1000 (**mean**/min/max) | #Matrix Mults/1000 (**mean**/min/max) | #Matrix Exps/1000 (**mean**/min/max) |
|---|---|---|---|---|---|---|
| 1 | conc. | **0.9999**/0.9999/1.0000 | **0.11**/0.02/0.26 | **1.43**/1.23/1.68 | **27**/23/31 | —none— |
|   | seq.  | **0.9999**/0.9999/0.9999 | **0.10**/0.04/0.22 | **6.17**/2.70/10.92 | **86**/38/152 | **6.17**/2.70/10.92 |
| 2 | conc. | **0.9999**/0.9999/0.9999 | **0.05**/0.03/0.10 | **2.10**/1.64/2.52 | **39**/31/47 | — |
|   | seq.  | **0.9999**/0.9999/0.9999 | **0.08**/0.05/0.09 | **5.74**/4.04/6.96 | **80**/56/97 | **5.74**/4.04/6.96 |
| 3 | conc. | **0.9999**/0.9999/1.0000 | **0.09**/0.08/0.13 | **4.49**/3.71/5.50 | **83**/68/102 | — |
|   | seq.  | **0.9999**/0.9999/0.9999 | **0.08**/0.03/0.15 | **5.13**/2.05/8.06 | **72**/29/113 | **5.13**/2.05/8.06 |
| 4 | conc. | **0.9999**/0.9999/1.0000 | **0.03**/0.02/0.04 | **1.60**/1.34/1.98 | **29**/24/37 | — |
|   | seq.  | **0.9999**/0.9999/0.9999 | **0.03**/0.01/0.04 | **1.89**/1.02/2.75 | **26**/14/38 | **1.89**/1.02/2.75 |
| 5 | conc. | **0.9877**/0.9322/0.9990 | **44**/24/67 | **364**/361/368 | **9828**/9759/9947 | — |
|   | seq.  | **0.9973**/0.9918/0.9986 | **34**/22/61 | **1976**/1320/3292 | **35542**/23729/59209 | **1976**/1320/3292 |
| 6 | conc. | **0.9999**/0.9999/0.9999 | **1.87**/0.73/3.61 | **24**/18/40 | **650**/473/1074 | — |
|   | seq.  | **0.9999**/0.9999/0.9999 | **5.34**/1.15/19.60 | **310**/68/1143 | **5574**/1216/20554 | **310**/68/1143 |
| 7 | conc. | **0.9958**/0.9808/0.9999 | **29.82**/5.74/69.99 | **244**/69/390 | **8552**/2411/13634 | — |
|   | seq.  | **0.9945**/0.9825/0.9999 | **123**/39/244 | **3978**/1242/7749 | **87443**/27313/170360 | **3978**/1242/7749 |
| 8 | conc. | **0.9999**/0.9999/0.9999 | **5.39**/2.08/20.83 | **49**/29/198 | **1697**/995/6925 | — |
|   | seq.  | **0.9999**/0.9999/0.9999 | **15.19**/2.66/67.58 | **500**/89/2223 | **11002**/1953/48876 | **500**/89/2223 |
| 9 | conc. | **0.9999**/0.9999/0.9999 | **1.21**/0.69/1.68 | **14.94**/8.19/21.06 | **521**/285/735 | — |
|   | seq.  | **0.9999**/0.9999/0.9999 | **4.90**/1.91/7.99 | **161**/63/259 | **3536**/1375/5696 | **161**/63/259 |
| 10 | conc. | **0.9999**/0.9997/0.9999 | **313**/21/1753 | **402**/94/904 | **17284**/4052/38874 | — |
|    | seq.  | **0.9999**/0.9999/0.9999 | **1059**/122/5497 | **4508**/1312/23582 | **117178**/34116/612977 | **4508**/1312/23582 |
| 11 | conc. | **0.9999**/0.9999/0.9999 | **153.76**/7.80/1104.29 | **144**/68/566 | **6182**/2890/24346 | — |
|    | seq.  | **0.9998**/0.9988/0.9999 | **1141**/86/9848 | **6990**/427/74922 | **181706**/11097/1947465 | **6990**/427/74922 |
| 12 | conc. | **0.9999**/0.9992/0.9999 | **76.27**/5.76/948.99 | **70**/22/194 | **3017**/936/8331 | — |
|    | seq.  | **0.9997**/0.9970/0.9999 | **1108**/39/5566 | **5054**/245/19200 | **131246**/6360/498598 | **5054**/245/19200 |
| 13 | conc. | **0.9844**/0.9102/0.9999 | **13.91**/2.49/75.62 | **120**/27/398 | **4189**/950/13926 | — |
|    | seq.  | **0.9759**/0.9373/0.9999 | **128.25**/6.64/454.86 | **4174**/215/15103 | **91773**/4719/332029 | **4174**/215/15103 |
| 14 | conc. | **0.9973**/0.9867/0.9999 | **7.47**/2.06/14.20 | **49**/29/80 | **1720**/1000/2801 | — |
|    | seq.  | **0.9999**/0.9999/0.9999 | **6.58**/1.77/16.61 | **219**/59/547 | **4813**/1292/12033 | **219**/59/547 |
| 15 | conc. | **0.9999**/0.9999/1.0000 | **0.05**/0.02/0.10 | **0.71**/0.52/0.92 | **7.49**/5.35/9.77 | — |
|    | seq.  | **0.9999**/0.9999/1.0000 | **0.02**/0.01/0.04 | **1.76**/0.72/3.28 | **17.53**/7.16/32.64 | **1.76**/0.72/3.28 |
| 16 | conc. | **0.9999**/0.9999/1.0000 | **0.04**/0.01/0.07 | **0.96**/0.77/1.34 | **9.99**/7.83/14.19 | — |
|    | seq.  | **0.9999**/0.9999/1.0000 | **0.01**/0.01/0.02 | **0.67**/0.51/1.54 | **6.64**/5.10/15.31 | **0.67**/0.51/1.54 |
| 17 | conc. | **0.9999**/0.9999/0.9999 | **531**/58/1443 | **1224**/1032/1551 | **13454**/11344/17054 | — |
|    | seq.  | **0.9999**/0.9999/0.9999 | **2284**/1054/3898 | **16774**/11490/27294 | **167710**/114877/272885 | **16774**/11490/27294 |
| 18 | conc. | **0.9999**/0.9999/0.9999 | **157**/26/754 | **574**/530/655 | **6300**/5821/7196 | — |
|    | seq.  | **0.9999**/0.9999/0.9999 | **386**/175/690 | **2953**/2434/3985 | **29524**/24335/39842 | **2953**/2434/3985 |
| 19 | conc. | **0.9999**/0.9999/0.9999 | **105**/16/335 | **166**/141/186 | **3807**/3244/4273 | — |
|    | seq.  | **0.9999**/0.9999/0.9999 | **143**/64/328 | **1130**/996/1465 | **18064**/15925/23433 | **1130**/996/1465 |
| 20 | conc. | **0.9999**/0.9999/0.9999 | **0.53**/0.12/1.14 | **5.15**/4.16/6.91 | **56**/45/76 | — |
|    | seq.  | **0.9999**/0.9999/0.9999 | **0.45**/0.23/0.77 | **30**/16/51 | **302**/160/511 | **30**/16/51 |
| 21 | conc. | **0.9999**/0.9999/0.9999 | **0.49**/0.36/0.89 | **9.53**/9.09/10.37 | **179**/171/195 | — |
|    | seq.  | **0.9999**/0.9999/0.9999 | **1.39**/0.79/3.80 | **39**/29/57 | **551**/410/804 | **39**/29/57 |
| 22 | conc. | **0.9999**/0.9999/0.9999 | **5.34**/2.39/8.03 | **131**/93/193 | **1444**/1026/2128 | — |
|    | seq.  | **0.9991**/0.9983/0.9999 | **108**/59/386 | **4702**/3317/6780 | **46924**/33106/67669 | **4702**/3317/6780 |
| 23 | conc. | **0.9999**/0.9999/0.9999 | **2.26**/0.42/9.57 | **38**/13/83 | **420**/148/913 | — |
|    | seq.  | **0.9951**/0.9797/0.9995 | **37.38**/9.40/97.52 | **2991**/744/7163 | **29786**/7408/71343 | **2991**/744/7163 |

# Bibliography

[1] W. S. Levine (ed). *The Control Handbook*. CRC Press, New York, 1996.

[2] M. Nielsen and I. Chuang. *Quantum Computation and Quantum Information*. Cambridge University Press, Cambridge, 2000.

[3] K. Resch, M. Lindenthal, B. Blauensteiner, H. Böhm, A. Fedrizzi, C. Kurtsiefer, A. Poppe, T. Schmitt-Manderbach, M. Taraba, R. Ursin, P. Walther, H. Weier, H. Weinfurter, and A. Zeilinger. Distributing entanglement and single photons through an intra-city, free-space quantum channel. *Opt. Express*, 13:202–209, 2005.

[4] R. P. Feynman. Simulating physics with computers. *Int. J. Theo. Phys.*, 21:467–488, 1982.

[5] D. Deutsch. The Church-Turing principle and the universal quantum computer. *P. R. Soc. London*, A 400:97–117, 1985.

[6] D. Deutsch and R. Jozsa. Rapid solutions of problems by quantum computation. *P. R. Soc. London*, A 439:553–558, 1992.

[7] P. W. Shor. Polynomial-time algorithms for prime factorization and discrete logarithms on a quantum computer. *SIAM J. Comput.*, 26:1484–1509, 1997.

[8] R. Marx, A. F. Fahmy, John M. Myers, W. Bermel, and S. J. Glaser. Approaching five-bit NMR quantum computing. *Phys. Rev. A*, 62:012310, 2000.

[9] T. Yamamoto, Yu Pashkin, O. Astafiev, Y. Nakamura, and J. S. Tsai. Demonstration of conditional gate operation using superconducting charge qubits. *Nature*, 425:941–944, 2003.

[10] B. E. Kane. A silicon-based nuclear spin quantum computer. *Nature*, 393:133–137, 1998.

[11] I. Chuang and Y. Yamamoto. Simple quantum computer. *Phys. Rev. A*, 52:3489–3496, 1995.

BIBLIOGRAPHY

[12] A. Imamoglu, D. D. Awschalom, G. Burkard, D. P. DiVincenzo, D. Loss, M. Sherwin, and A. Small. Quantum information processing using quantum dot spins and cavity qed. *Phys. Rev. Lett.*, 83:4204–4207, 1999.

[13] J. E. Mooij, T. Orlando, P., L. Levitov, Lin Tian, Van, Caspar H. der Wal, and Seth Lloyd. Josephson persistent-current qubit. *Science*, 285:1036–1039, 1999.

[14] J. I. Cirac and P. Zoller. Quantum computations with cold trapped ions. *Phys. Rev. Lett.*, 74:4091–4094, 1995.

[15] A. E. Bryson and Y.-C. Ho. *Applied Optimal Control*. Blaisdell Pub. Co. Waltham, Mass., 1969.

[16] D. E. Kirk. *Optimal Control Theory: An Introduction*. Dover, 1970.

[17] H. J. Sussmann and J. C. Willems. 300 years of optimal control: from the brachystochrone to the maximum principle. *IEEE Contr. Syst. Mag.*, 17:32–44, 1997.

[18] L. S. Pontryagin, V. G. Boltyanskii, R. V. Gamkrelidze, and E. Mishchenko. *The Mathematical Theory of Optimal Processes*. Interscience Publishers, 1962.

[19] R. Bellman. On the theory of dynamic programming. In *Proceedings of the National Academy of Sciences*, volume 38, pages 716–719, 1952.

[20] N. Khaneja and S. J. Glaser. Cartan decomposition of $su(2^n)$ and control of spin systems. *Chem. Phys.*, 267:11–23, 2001.

[21] M. Lapert, Y. Zhang, M. Braun, S. J. Glaser, and D. Sugny. Singular extremals for the time-optimal control of dissipative spin 1/2 particles. *Phys. Rev. Lett.*, 104:083001, 2010.

[22] A. Spörl, T. Schulte-Herbrüggen, S. J. Glaser, V. Bergholm, M. J. Storcz, J. Ferber, and F. K. Wilhelm. Optimal control of coupled josephson qubits. *Phys. Rev. A*, 75:012302, 2007.

[23] R. Fisher, F. Helmer, S. J. Glaser, F. Marquardt, and T. Schulte-Herbrüggen. Optimal control of circuit quantum electrodynamics in one and two dimensions. *Phys. Rev. B*, 81:085328, 2010.

[24] R. Nigmatullin and S. G. Schirmer. Implementation of fault-tolerant quantum logic gates via optimal control. *New J. Phys.*, 11:105032, 2009.

[25] J. Nocedal and S. J. Wright. *Numerical Optimization*. Springer, 2000.

[26] J. F. Cornwell. *Group Theory in Physics*. Volume 2, Academic Press, London, 1984.

[27] B. L. Van der Waerden. *Algebra I*. Springer-Verlag, Berlin, 1971.

[28] G. Pickert. *Einführung in die Höhere Algebra*. Vandenhoeck & Ruprecht, Göttingen, 1951.

[29] C. Chevalley. *Fundamental Concepts of Algebra*. Academic Press, New York, 1956.

[30] J. Hilgert and K.H. Neeb. *Lie-Gruppen und Lie-Algebren*. Springer-Verlag, Berlin, 1991.

[31] M. Nakahara. *Geometry, Topology and Physics*. Taylor & Francis Group, Boca Raton, 2003.

[32] V. Ramakrishna and H. Rabitz. Relation between quantum computing and quantum controllability. *Phys. Rev. A*, 54:1715–1716, 1995.

[33] S. J. Glaser, T. Schulte-Herbrüggen, M. Sieveking, O. Schedletzky, N. C. Nielsen, O. W. Sørensen, and C. Griesinger. Unitary control in quantum ensembles: maximising signal intensity in coherent spectroscopy. *Science*, 280:421–424, 1998.

[34] S. Bose. Quantum communication through spin-chain dynamics: an introductory overview. *Contemp. Phys.*, 48:13–30, 2007.

[35] R. van Meter, K. Nemoto, and W. J. Munroe. Quantum communication through spin-chain dynamics: an introductory overview. *IEEE T. Comput.*, 56:1643–1653, 2007.

[36] S. Lloyd. A potentially realizable quantum computer. *Science*, 261:1569–1571, 1993.

[37] J. Fitzsimons and J. Twamley. Globally controlled wires for perfect qubit transport, mirroring, and computing. *Phys. Rev. Lett.*, 99:090502, 2006.

[38] S. C. Benjamin. Schemes for parallel quantum computing without local control of qubits. *Phys. Rev. A*, 61:020301, 2000.

[39] S. Bose. Quantum communication through an unmodulated spin chain. *Phys. Rev. Lett.*, 91:207901, 2003.

[40] C. Christandl, N. Datta, A. Ekert, and A. J. Landahl. Perfect state transfer in quantum spin networks. *Phys. Rev. Lett.*, 92:187902, 2004.

[41] J. Eisert, M. B. Plenio, S. Bose, and J. Hartley. Towards quantum entanglement in nano-electromechanical devices. *Phys. Rev. Lett.*, 93:190402, 2004.

BIBLIOGRAPHY

[42] D. Burgarth and S. Bose. Perfect quantum state transfer with randomly coupled quantum chains. *New J. Phys.*, 7:135, 2005.

[43] S. G. Schirmer, I. C. H. Pullen, and P. J. Pemberton-Ross. Global controllability with a single local actuator. *Phys. Rev. A*, 78:030501, 2008.

[44] H. Sussmann and V. Jurdjevic. Controllability of nonlinear systems. *J. Diff. Equat.*, 12:95–116, 1972.

[45] V. Jurdjevic and H. Sussmann. Control systems on lie groups. *J. Diff. Equat.*, 12:313–329, 1972.

[46] T. Schulte-Herbrüggen. *Aspects and Prospects of High-Resolution NMR*. PhD Thesis, Diss-ETH 12752, Zürich, 1998.

[47] F. Albertini and D. D'Alessandro. The lie algebra structure and controllability of spin systems. *Lin. Alg. Appl.*, 350:213–235, 2002.

[48] C. Altafini. Controllability of quantum mechanical systems by root space decomposition of $su(N)$. *J. Math. Phys.*, 43:2051–2062, 2002.

[49] D. Burgarth, S. Bose, C. Bruder, and V. Giovanetti. Local controllability of quantum networks. 2008. e-print: arXiv:0805.3975v2 [quant-ph].

[50] N. Khaneja, T. Reiss, C. Kehlet, T. Schulte-Herbrüggen, and S. J. Glaser. Optimal control of coupled spin dynamics: Design of NMR pulse sequences by gradient ascent algorithms. *J. Magn. Reson.*, 172:296–305, 2005.

[51] F. Albertini and D. D'Alessandro. Notions of controllability for bilinear multilevel quantum systems. *IEEE T. Automat. Contr.*, 48:1399–1403, 2003.

[52] E. B. Lee and L. Markus. *Foundations of Optimal Control Theory*. Wiley, New York, 1967.

[53] R. W. Brockett. System theory on group manifolds and coset spaces. *SIAM J. Control*, 10:265–284, 1972.

[54] V. Jurdjevic. *Geometric Control Theory*. Cambridge University Press, Cambridge, 1997.

[55] V. Ramakrishna, M. Salapaka, M. Daleh, H. Rabitz, and A. Peirce. Controllability of molecular systems. *Phys. Rev. A*, 51:960–966, 1995.

BIBLIOGRAPHY

[56] S. G. Schirmer, H. Fu, and A. I. Solomon. Complete controllability of quantum systems. *Phys. Rev. A*, 63:063410, 2001.

[57] R. A. Horn and C. R. Johnson. *Topics in Matrix Analysis*. Cambridge University Press, Cambridge, 1991.

[58] The GAP Group. *GAP – Groups, Algorithms, and Programming, Version 4.4.10*, 2008. (http://www.gap-system.org).

[59] T. Schulte-Herbrüggen, A. K. Spörl, N. Khaneja, and S. J. Glaser. Optimal control-based efficient synthesis of building blocks of quantum algorithms: A perspective from network complexity towards time complexity. *Phys. Rev. A*, 72:042331, 2005.

[60] T. Schulte-Herbrüggen, Z. Mádi, O. W. Sørensen, and R. R. Ernst. Reduction of multiplet complexity in COSY-type NMR spectra: bilinear and planar COSY experiments. *Molec. Phys.*, 72:847–871, 1991.

[61] D. Burgarth, K. Maruyama, S. Montangero, T. Calarco, F. Noi, and M. Plenio. Scalable quantum computation via local control of only two qubits. 2009. e-print: arXiv:0905.3373v2 [quant-ph].

[62] A. Kay and P. Pemberton-Ross. Computation on spin chains with limited access. 2009. e-print: arXiv:0905.4070v3 [quant-ph].

[63] R. Gilmore. *Lie Groups, Lie Algebras, and some of their Applications*. Krieger Publishing Company, Florida, 1994.

[64] D. D'Alessandro. *Introduction to Quantum Control and Dynamics*. Chapman & Hall/CRC, Boca Raton, 2008.

[65] G. Turinici and H. Rabitz. Wavefunction controllability for finite-dimensional bilinear quantum systems. *J. Phys. A*, 36:2565–2576, 2003.

[66] A. W. Knapp. *Lie Groups beyond an Introduction*. Birkhäuser, Boston, 2nd edition, 2002.

[67] R. Zeier and T. Schulte-Herbrüggen. In preparation.

[68] R. F. Werner and A. S. Holevo. Counterexample to an additivity conjecture for output purity of quantum channels. *J. Math. Phys.*, 43:4353–4357, 2002.

[69] M. Choi. Completely positive linear maps on complex matrices. *Lin. Alg. Appl.*, 12:95–100, 1975.

BIBLIOGRAPHY

[70] A. S. Holevo. Private communication, 2007.

[71] C. Mendl and M. Wolf. Unital quantum channels - convex structure and revivals of Birkhoff's theorem. *Comm. Math. Phys.*, 289:1057–1086, 2009.

[72] C. King. Additivity for a class of unital qubit channels. 2001. e-print: arXiv:quant-ph/0103156v2.

[73] C. King. The capacity of the quantum depolarizing channel. *IEEE T Inform Theory*, 49:221–229, 2003.

[74] M. Gregoratti and R. F. Werner. Quantum lost and found. *J. Mod. Opt.*, 50:915–933, 2003.

[75] P. Hayden and A. Winter. Counterexamples to the maximal p-norm multiplicativity conjecture for all p > 1. *Comm. Math. Phys.*, 284:263–280, 2008.

[76] M. B. Hastings. Superadditivity of communication capacity using entangled inputs. *Nature Physics*, 5:255–257, 2008.

[77] V. F. Krotov and I. N. Feldman. Iteration method of solving the problems of optimal control. *Eng. Cybern.*, 21:123–130, 1983. Russian original: *Izv. Akad. Nauk. SSSR Tekh. Kibern.* **52** (1983), 162–167.

[78] A. I. Konnov and V. F. Krotov. On the global methods of successive improvement of controllable processes. *Automat. Rem. Contr.*, 60:1427, 1999. Russian original: *Avtom. Telemekh.* **1999**, 77–88.

[79] V. F. Krotov. *Global Methods in Optimal Control*. Marcel Dekker, New York, 1996.

[80] N. I. Gershenzon, K. Kobzar, B. Luy, S. J. Glaser, and T. E. Skinner. Optimal control design of excitation pulses that accommodate relaxation. *J. Magn. Reson.*, 188:330–336, 2007.

[81] S. Stepanenko and B. Engels. Gradient tabu search. *J. Comput. Chem.*, 28:601–611, 2006.

[82] S. Stepanenko and B. Engels. New tabu search based global optimization methods: outline of algorithms and study of efficiency. *J. Comput. Chem.*, 29:768–780, 2007.

[83] R. Karplus and J. Schwinger. A note on saturation in microwave spectroscopy. *Phys. Rev.*, 73:1020–1026, 1948.

BIBLIOGRAPHY

[84] K. Aizu. Parameter differentiation of quantum-mechanical linear operators. *J. Math. Phys.*, 4:762–775, 1963.

[85] I. Kuprov and C. T. Rodgers. Derivatives of spin dynamics simulations. *J. Chem. Phys.*, 131:234108, 2009.

[86] D. Suter and T. S. Mahesh. Spins as qubits: Quantum information processing by nuclear magnetic resonance. *J. Chem. Phys.*, 128:052206, 2008.

[87] T. E. Skinner, K. Kobzar, B. Luy, R. Bendall, W. Bermel, N. Khaneja, and S. J. Glaser. Optimal control design of constant amplitude phase-modulated pulses: application to calibration-free broadband excitation. *J. Magn. Reson.*, 179:241–249, 2006.

[88] J. L. Neves, B. Heitmann, N. Khaneja, and S. J. Glaser. Heteronuclear decoupling by optimal tracking. *J. Magn. Reson.*, 201:7–17, 2009.

[89] A. Mordecai. *Nonlinear Programming: Analysis and Methods*. Dover Publications, 2003.

[90] M. R. Hestenes and E. Stiefel. Methods of conjugate gradients for solving linear systems. *J. Res. Natl. Bur. Stand.*, 49:409–436, 1952.

[91] I. Kuprov, N. Wagner-Rundell, and P.J. Hore. Polynomially scaling spin dynamics simulation algorithm based on adaptive state-space restriction. *J. Magn. Reson.*, 189:241–250, 2007.

[92] I. Kuprov. Polynomially scaling spin dynamics II: Further state-space compression using Krylov subspace techniques and zero track elimination. *J. Magn. Reson.*, 195:45–51, 2008.

[93] M. Schmidt. minFunc software package for Matlab. (http://www.cs.ubc.ca/~schmidtm/Software/minFunc.html), 2010.

[94] H. Wunderlich, C. Wunderlich, K. Singer, and F. Schmidt-Kaler. Two-dimensional cluster-state preparation with linear ion traps. *Phys. Rev. A*, 79:052324, 2009.

[95] P. Neumann, N. Mizuochi, F. Rempp, P. Hemmer, H. Watanabe, S. Yamasaki, V. Jacques, T. Gaebel, F. Jelezko, and J. Wrachtrup. Multipartite entanglement among single spins in diamond. *Science*, 320:1326–1329, 2008.

[96] U. Sander and T. Schulte-Herbrüggen. Controllability and observability of multi-spin systems: Constraints by symmetry and by relaxation. 2009. e-print: arXiv:0904.4654v2 [quant-ph].

[97] T. O. Levante, T. Bremi, and R. R. Ernst. Pulse-sequence optimization with analytical derivatives. Application to deuterium decoupling in oriented phases. *J. Magn. Reson. A*, 121:167–177, 1996.

[98] R. H. Byrd, M. E. Hribar, and J. Nocedal. An interior point algorithm for large scale nonlinear programming. *SIAM J. Optim.*, 9:877–900, 1999.

[99] R. A. Waltz, J. L. Morales, J. Nocedal, and D. Orban. An interior algorithm for nonlinear optimization that combines line search and trust region steps. *Math. Program.*, 107:391–408, 2006.

Die VDM Verlagsservicegesellschaft sucht für wissenschaftliche Verlage abgeschlossene und herausragende

## Dissertationen, Habilitationen, Diplomarbeiten, Master Theses, Magisterarbeiten usw.

### für die kostenlose Publikation als Fachbuch.

Sie verfügen über eine Arbeit, die hohen inhaltlichen und formalen Ansprüchen genügt, und haben Interesse an einer honorarvergüteten Publikation?

Dann senden Sie bitte erste Informationen über sich und Ihre Arbeit per Email an *info@vdm-vsg.de*.

**Sie erhalten kurzfristig unser Feedback!**

VDM Verlagsservicegesellschaft mbH
Dudweiler Landstr. 99            Telefon  +49 681 3720 174
D - 66123 Saarbrücken         Fax         +49 681 3720 1749
**www.vdm-vsg.de**

Die VDM Verlagsservicegesellschaft mbH vertritt

Printed by Books on Demand GmbH, Norderstedt / Germany